WITHDRAWN

Procedures and Practices in Activated Sludge Process Control

ACTIVATED SLUDGE PROCESS CONTROL SERIES

Edited by Robert M. Arthur
and the staff of Arthur Technology

Volume 1. NEW CONCEPTS AND PRACTICES IN ACTIVATED SLUDGE PROCESS CONTROL
R. M. Arthur

Volume 2. APPLICATION OF ON-LINE ANALYTICAL INSTRUMENTATION TO PROCESS CONTROL
R. M. Arthur, Editor

Volume 3. PROCEDURES AND PRACTICES IN ACTIVATED SLUDGE PROCESS CONTROL
R. M. Arthur, Editor

Procedures and Practices in Activated Sludge Process Control

Edited by
Robert M. Arthur, PhD

ARTHUR
TECHNOLOGY

In co-publication with

BUTTERWORTH PUBLISHERS
Boston • London
Sydney • Wellington • Durban • Toronto

An Ann Arbor Science Book

Ann Arbor Science is an imprint of Butterworth Publishers

Copyright © 1983 by Arthur Technology, Inc., 548 Prairie Road, P.O. Box 1222, Fond du Lac, Wisconsin 54935

No part of this publication may be reproduced, stored in a retrieval system, or transmitted, in any form or by any means, electronic, mechanical, photocopying, recording, or otherwise, without the prior written permission of the publisher.

Library of Congress Catalog Card Number 83-071517
ISBN 0-250-40630-6

Published by Butterworth Publishers
10 Tower Office Park
Woburn, MA 01801

Printed in the United States of America

10 9 8 7 6 5 4 3 2 1

This book is dedicated
to those who made the
Second Annual Conference on
Activated Sludge a success—
the speakers and the attendees.

Robert M. Arthur is founder and President of Arthur Technology, a consulting, research and development, testing and training firm in Wisconsin that specializes in wastewater treatment plant operation and control. A registered Professional Engineer, he received his PhD in 1963 from the University of Iowa. In addition to his work at Arthur Technology, he taught for several years at Rose Hulman Institute of Technology where he founded, and was chairman of, the Biological Engineering Department. He has also been active in teaching courses in a two-year associate degree program in water and wastewater technology which he initiated at a Wisconsin technical institute. Dr. Arthur serves on the Standard Methods Committee of the Water Pollution Control Federation and has been chairman of the Instrumentation Subcommittee of the Water Pollution Control Federation. Recipient of many national awards, he holds six patents, is the author of over 30 papers and was the 1981 winner of the Kermit Fischer Environmental Award of the Instrument Society of America. He is the editor of the Activated Sludge Process Control Series, co-published by Arthur Technology and Ann Arbor Science.

CONTENTS

Preface xi

1. Correlation of Process Control Strategies with
   Effluent BOD$_5$ and SS 1
   Dr. Edward D. Schroeder

2. Process Control Strategies 25
   Dr. Robert M. Arthur

3. Simplified Control Strategies—
   Letting the Operator Operate 47
   Paul Klopping

4. Understanding the Application of Analytical Data
   to Process Control 53
   Thomas J. Kutcher and Gary E. Ettel

5. Data Management for Process Control 65
   Owen K. Boe

6. Computerization—Can it Work for You? 77
   Robert A. Davis

7. The Computer as a Tool for Activated Sludge 87
   Robert G. Skrentner

8. Relating BOD$_5$ with On-Line Oxygen Uptake Rate
   Measurements Using Automatic Respirometers in
   View of Process Monitoring and Control 113
   Dr. Normand Therien and Ferhat Ilhan

9. Design of Activated Sludge Facilities for Operability and Maintainability 131
   Vernon T. Stack and Edward L. Gillette, Jr.

10. Good Process Control: Planning and Implementation 153
    Ed R. Fioroni and Walter D. Frais

11. Startup Critique of Sartell 3 173
    David L. Keller

12. Round Table Discussion 181

Appendix: Question and Answer Sessions 209

Index 223

PREFACE

The Second Annual Conference on Activated Sludge Process Control was planned to duplicate the successful format of the 1981 conference. A theme for the meeting was selected first, then hand-picked speakers were selected to present the papers. Topics for each paper were assigned to each speaker based on his particular expertise. The order of presentation was scheduled to logically introduce and develop the general theme of the conference.

This year's theme, "Procedures and Practices in Activated Sludge Process Control," was selected by reviewing theme suggestions made by attendees of the First Annual Conference. The theme reflects an interest in practical "hands-on" experience in process control. Past attendees wanted more information on control schemes and strategies, and computer control, and wanted speakers with actual plant experience. The topics and speakers for this year's conference certainly satisfy this request.

The success of the first and second conferences is ample evidence of growing interest in process control of activated sludge. The conference proceedings should become an excellent reference work on this important topic.

Robert M. Arthur, PhD

CHAPTER 1

CORRELATION OF PROCESS CONTROL STRATEGIES WITH EFFLUENT BOD AND SUSPENDED SOLIDS

Dr. Edward D. Schroeder
Department of Civil Engineering
University of California—Davis
Davis, California

INTRODUCTION

Biological wastewater treatment processes are expected, by designers, regulators and operators, to behave in a consistent manner. This means that any variation in performance is a bit surprising, despite the fact that no one has ever seen a biological wastewater treatment plant operating at true steady state. There are a number of reasons why steady state is an unachievable goal. First and foremost is the variation in input. Large variations occur in all wastewater streams, although the cycling of the variations may be quite different in municipal and industrial wastewaters. Some sequencing in population dominance among the species making up the microbial culture can be expected also. Those who have performed regular microscopic examinations of activated sludge floc or trickling filter slimes are well aware of the procession of species of the larger organisms such as protozoans and rotifers, as well as the changes in floc structure that accompany phenomena such as bulking. Temperature is a third major factor in natural variation in process performance. Reaction rates are quite sensitive to temperature, although stoichiometry changes very little within the operating temperature range.

OBJECTIVES OF PROCESS CONTROL

Because biological wastewater treatment processes cannot be expected to operate at steady state, and therfore cannot produce the design effluent characteristics on a constant basis, the introduction of process control is desirable. Process control is a loosely defined term that can have a number of meanings. Here the meaning is restricted to actions that directly affect performance. Thus, maintaining dissolved oxygen concentrations above 2.0 g/m³ might require a control procedure, but it would not be process control in cases. Chemical addition for the purpose of producing a low-turbidity effluent would be directly related to performance, and the control of chemical addition would be a form of process control.

In the activated sludge process the principal objectives of process control are: (1) damping out of input variation, (2) minimization of effluent variation and (3) prevention and recovery from process upsets. Control of these three factors is directly related to process reliability in terms of meeting discharge requirements. Control strategies have been developed for all three objectives, but results have been mixed for a variety of reasons.

Development of process control techniques for trickling filters has been much slower than in the case of activated sludge systems. Active process control of trickling filters is nonexistent in practice, and design engineers generally do not attempt to include operator-controllable features other than recycle in the systems. For this reason discussion of control strategies for trickling filters must be limited to suggestions for future work.

COMMON ACTIVATED SLUDGE CONTROL STRATEGIES

Control strategies in common use include:

- sludge age/wasting
- loading rate
- flow equalization
- process configuration
- recycle
- dissolved oxygen
- chemical addition

Sludge Age Control

Sludge age, often called solids retention time (SRT) or mean cell residence time (MCRT), is a very useful parameter because it can be related to growth rate and organic removal rate. The derivation of these relationships and the stoichiometric coefficients is now standard in textbooks [1–3] and will not be done here, but the result is:

(a) For a continuous flow stirred tank reactor (CFSTR) at steady state

$$r_g = SRT^{-1} \tag{1}$$

$$r_o = Y_g r_g \tag{2}$$

where
r_g = specific growth rate (t^{-1})
r_o = specific organic removal rate (t^{-1})
Y_g = net solids yield

(b) For an ideal plug flow reactor (PFR) at steady state the expressions analogous to Equations 1 and 2 are based on mass balances around a differential section.

$$\frac{dX}{d\theta} = r_g X \tag{3}$$

$$\frac{dC}{d\theta} = r_o X = -Y_g r_g X \tag{4}$$

where
X = cell mass concentration or a related variable such as mixed liquor suspended solids (MLSS), (g/m^3)
C = organic concentration, preferably stated as ultimate BOD (g/m^3)

The most common representation of the growth rate is the modified Monod expression:

$$r_g = \frac{r_m C}{K+C} - k_d \tag{5}$$

where
r_m = maximum specific growth rate (t^{-1})
k_d = endogenous respiration rate (t^{-1})
K = saturation coefficient (g/m^3)

Inserting a reaction model such as Equation 5 into Equations 3 and 4 and solving the differential equations results in expressions relating effluent substrate concentration to average SRT. These equations are more complex than Equations 1 and 2 for CFSTR, but they are analogous. For both reactor configurations the SRT value and the effluent BOD concentration are directly related. The results are dependent on the boundary conditions used in the integration, but the values shown in Figure 1 are typical.

The important feature of the relationships is that SRT and effluent BOD are coupled. Because SRT is set by controlling sludge wasting, the

4 ACTIVATED SLUDGE PROCESS CONTROL

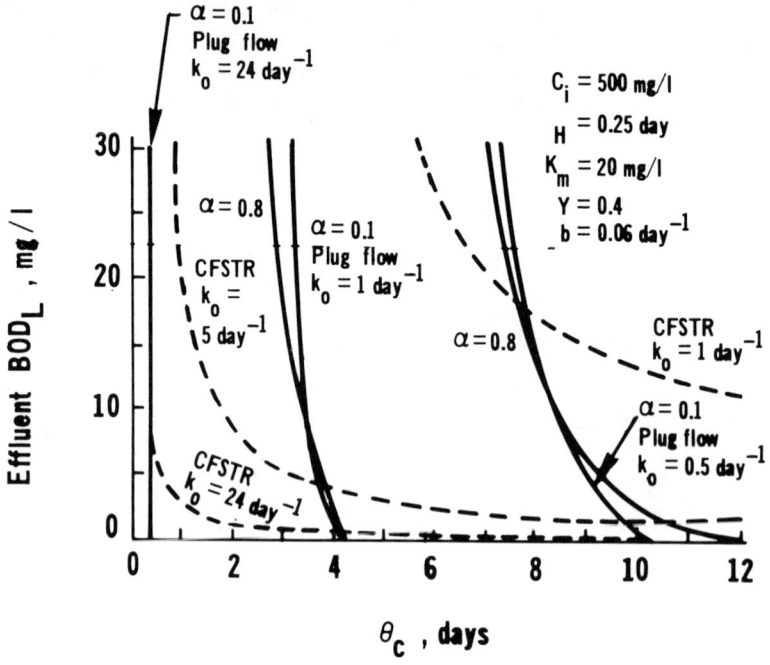

Figure 1. Relationship between reactor configuration and performance for steady-state conditions.

concept of SRT as a process control parameter developed. Real processes are neither ideal CFSTR or ideal PFR and therefore the actual control procedures must have some flexibility. As can be seen in Figure 1, however, the actual difference in the predicted effluent quality within the operating region (SRT > 3d) is relatively small. Thus an operator should be able to set a treatment plant at a chosen SRT and produce a steady effluent quality.

Flow Equalization

All treatment plant inflows vary with time. Municipal sewage flow typically varies by a factor of at least two over the course of a day, and by factors of four or more over the course of a year. Industrial flows tend to be intermittant except in large industries. Organic concentration often varies in the same manner as flow rate, which means that the organic input or loading rate has a greater variation. Because aeration capacity and sedimentation tank overflow rate are usually based on maximum daily flow or a similar value, there is good reason to minimize flow variation. Both factors can affect process performance and therefore flow equalization is generally desirable.

In fact, flow equalization is difficult to incorporate into process design because of potential for generating anaerobic conditions. Equalization capacity must be based on maximum wet weather flows. This means that the equalization tanks have long hydraulic retention times during low-flow periods. Low flow usually occurs during warm summer days when the potential for odors is greatest and for this reason very few treatment plants incorporate flow equalization. Where equalization is used aeration is provided in the tank and a considerable fraction of the influent BOD is removed. This may result in problems due to overdesign of the activated sludge system.

Loading Rate

The use of organic loading rate as a control parameter is very similar to the use of SRT because the two are stoichiometrically related. Organic loading rate is usually defined in terms of the food to microorganism ratio (F/M):

$$F/M = QC_i/VX \tag{6}$$

where
Q = volumetric flow rate (m^3/s)
C_i = influent BOD (g/m^3)
V = aeration tank volume (m^3)

Because the effluent BOD concentration is usually quite small, $C_i-C = C_i$ is an adequate approximation and the F/M is virtually identical with the specific removal rate. Operational control of the input term, QC_i, is generally not possible; tank volume is fixed in continuous flow processes and this means that MLSS concentration is the control variable, just as in SRT control. Thus the method of control as well as the concepts are identical.

One difference exists between F/M and SRT control procedures. Solids residence time control is directed at producing an effluent of a given quality. Thus the SRT value chosen is related to a kinetic model and an effluent requirement. The food to microorganism ratio is chosen on the basis of experience and values used are loosely defined as an operating region. For example, a general rule of thumb is that $0.3 < F/M < 0.75$ (based on BOD$_u$) is an operating region within which both a satisfactory effluent quality and a sludge that settles well will be produced. Values of

F/M > 0.75 often result in dispersed growth and turbid effluents, while values below 0.3 push the system into a nitrification mode of operation.

Operators using F/M control are expected to stay within a stated range by wasting control, and to modify the F/M in response to specific process changes such as seasonal industrial discharges, changes in temperature and filamentous bulking.

Process Configuration

Many activated sludge plants are designed to allow operation under a number of configurations. For example, a plant might be piped for operation as a conventional tapered aeration system, a step feed system or a contact stabilization process. In some cases the concept is to change configuration in response to influent characteristic changes, e.g., beginning of canning season. More often there is an intent to provide the operator with some flexibility in responding to performance changes. This latter reason demonstrates the state of understanding of the process quite well. In most cases if the process is upset (which usually means filamentous bulking) the configuration is changed in a blind hope that the upset will go away.

Recycle Ratio Variation

Some flexibility in setting recycle rate is necessary because settled sludge concentration varies from time to time. Most plants are designed to provide recycle ratios up to one. Performance of systems that are designed as nominal PFR is somewhat affected by recycle because the influent organic concentration is decreased, but CFSTR are unaffected. Hydraulic loading of the clarifier is affected by recycle in both configurations. Thus any increase in recycle ratio should have a negative effect on performance.

Dissolved Oxygen

Oxygen is required as an electron acceptor in the bioxidation reactions of activated sludge treatment, and like any other reactant it is a potential rate-limiting material. Evidence of rate limitation in activated sludge processes is contradictory. Limitations are often predicted based on theoretical models, but experimental results usually indicate no limitations when dissolved oxygen concentration is above 1.0 g/m^3 [1,4,7]. Heukelekian and Ingols [8] suggested that filamentous bulking was associated with low dissolved oxygen concentrations and Sezgin et al [9] recently reported similar findings. A relationship between low oxygen concentrations and filamentous bulking is probably an artifact of other limitations, however. This conclusion is based on the inconsistancy of the response. Many plants with dissolved oxygen concentrations set at values near 1.0 g/m^3 do not have bulking problems, while plants that are maintained at high concentration do bulk. Examples of the latter situation include pure oxygen processes. In addition, anoxic conditions have been shown to prevent bulking when coupled to high organic concentrations such as those occurring at the inlet of a conventional process [10, 11]. A more likely relationship is between bulking and low organic concentrations [12, 13].

Chemical Addition

Chemicals are added to activated sludge processes for four reasons:

1. to provide buffering
2. to provide excess amounts of required nutrients such as nitrogen
3. to control undesirable microorganisms
4. to improve clarification

Buffers and nutrients need to be provided in excess and therefore only effect performance indirectly. Coagulants need to be added under controlled conditions because of their expense and because an optimal value may exist. Thus coagulant addition can be a control variable. Control of undesirable microorganisms is usually accomplished by chlorination of the recycled activated sludge. Filamentous organisms appear to be more sensitive to chlorination than flocculant organisms, probably because the latter are protected by the polymers that make up most of the floc volume. Careful control of chlorine added, usually on a mass Cl_2/mass MLSS basis, results in chlorination rate becoming a control parameter in some systems.

PREDICTED AND ACTUAL PROCESS PERFORMANCE

The response of activated sludge processes to the available control strategies cannot be predicted in all cases. Some are either/or situations, as was noted in the case of buffer and nutrient addition and dissolved oxygen concentration. Recycled sludge chlorination to control filamentous bulking may be a control parameter, but the effects are not quantifiable at the present time because a predictive model for effluent suspended solids concentration does not exist. The effect of SRT, F/M and process configuration on performance can be predicted if it is assumed that instabilities do not occur.

Control Using SRT or F/M

A considerable amount of work has been done on the response of activated sludge processes to transient loadings. Most of the work, both experimental and theoretical, has been limited to organic loading rate transients because of the additional complexity of including hydraulic effects as well [14-16]. Response of systems originally at steady state to organic loading perturbations is very damped, as shown in Figure 2. A large change in influent concentration is required before a significant increase in effluent concentration occurs. Deflocculation is not a rapid response and floc formation appears to be more of an effect of average than point conditions [15]. This conclusion is supported by the work with batch systems of Irvine and co-workers [17-19] and Silverstein [20]. McLellan and Busch [21] studied combined hydraulic and organic transients and reported similar results.

A theoretical model with both varying hydraulic and organic loading was developed by Schroeder and Silverstein [22]. In-phase cyclic variations of flowrate and influent BOD concentration were used such that flowrate varied by a factor of two and BOD mass input rate varied by a factor of four. Effluent suspended solids were varied as a function of the ratio of actual to average flowrate. No attempt was made to impose a variable solids compactability and hence variable recycle rate, but as can be seen by comparing Figures 3 and 4 hydraulic effects are not major factors in the system response.

The responses to organic and hydraulic transients shown in Figures 3 and 4 included two control strategies based on SRT. Minimizing the effluent BOD variation requires increasing the removal rate. When only an organic transient occurs MLSS concentrations quickly increase and a sharply damped output results. Hydraulic transients result in a decrease in MLSS. The most appropriate response is to stop wasting to minimize the decrease and to maximize the rate of return to the normal cyclic variation range. Note that, considering the range in organic and hydraulic loading in each 24-hour cycle, a very small range of variation in effluent quality is predicted.

The principal conclusion that can be drawn from a review of both the experiemental and transient response research reported in the literature is that activated sludge processes are very slow to respond to hydraulic or organic perturbations. Changes do not occur suddenly and the response is generally very damped compared to the input signal. Variation in effluent quality that occurs in experimental systems operated at steady state is

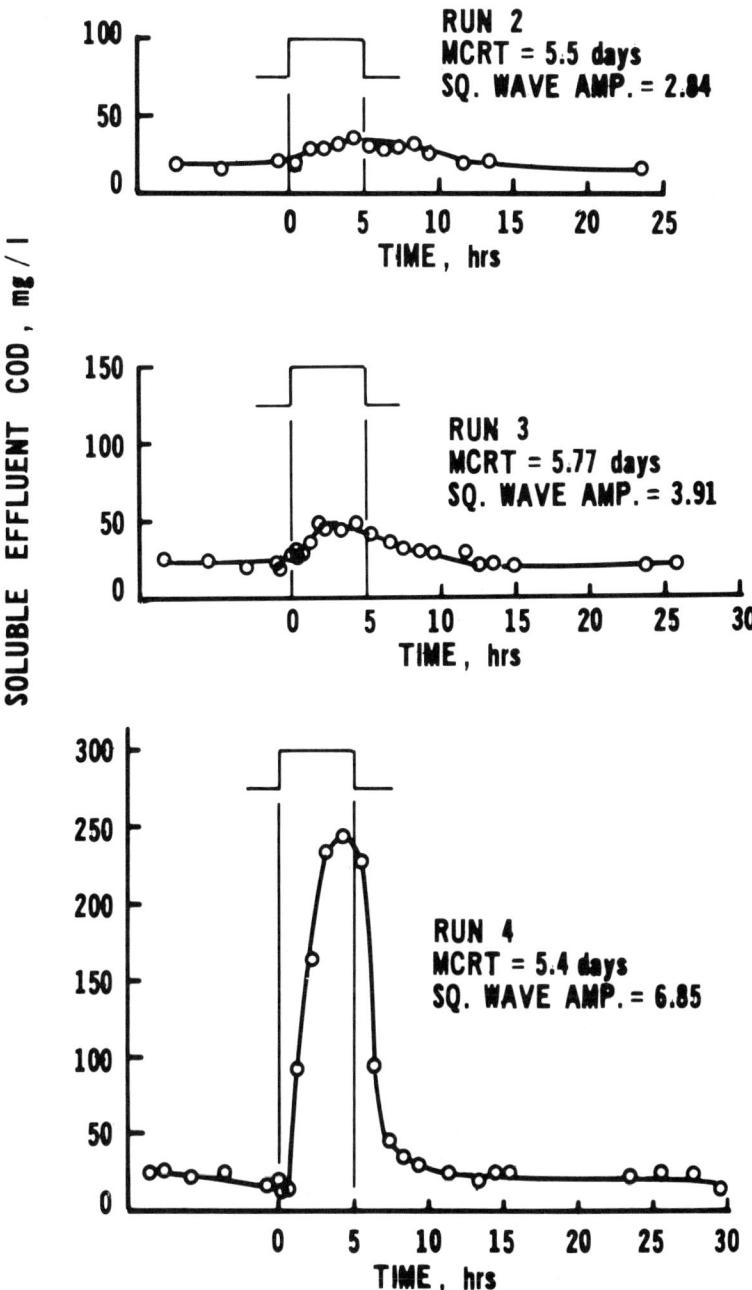

Figure 2. Response of CFSTR activated sludge processes to organic transients [15].

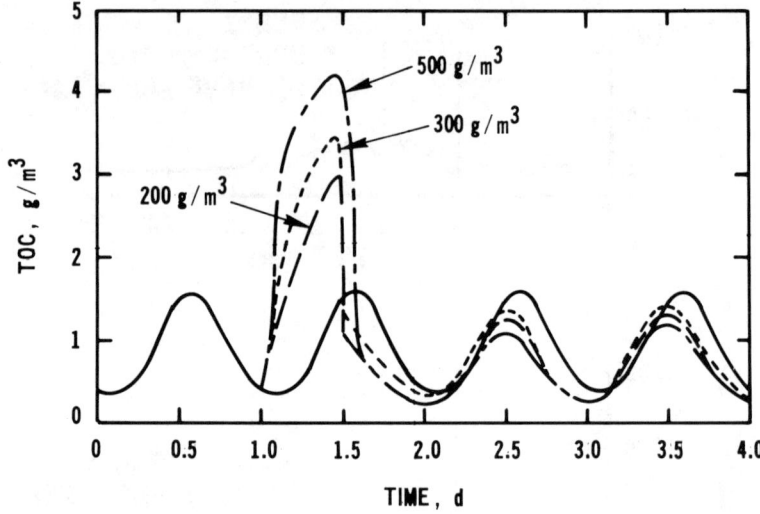

Figure 3. TOC response to step change in influent with no change in Hydraulic Loading

generally very small. It is difficult to determine how much of the variation results from analytical accuracy and how much results from natural succession.

Control By Change of Process Configuration

As in the case of response to transient loadings, changing process configuration will have little effect on predicted process performance. In the SRT range used with activated sludge processes, effluent quality predicted using kinetic modeling will be nearly independent of configuration as was shown in Figure 1. The major reason for changing process configuration is to respond to an upset, and whether or not this is a valid procedure is not known.

There have been a number of evaluations of activated sludge process performance on a statistical basis [23–30]. A recent report by Niku and Schroeder [30] evaluated the relationship of a large number of process variables and parameters to the variability of effluent BOD and SS. This work was an extension of a previous report [29] dealing with reliability-based design. In the study of the relationship between process variables and effluent quality, one calendar year of daily data, including process variable values, was available for 21 activated sludge treatment plants. Nine of the plants were conventional, two were completely mixed, seven

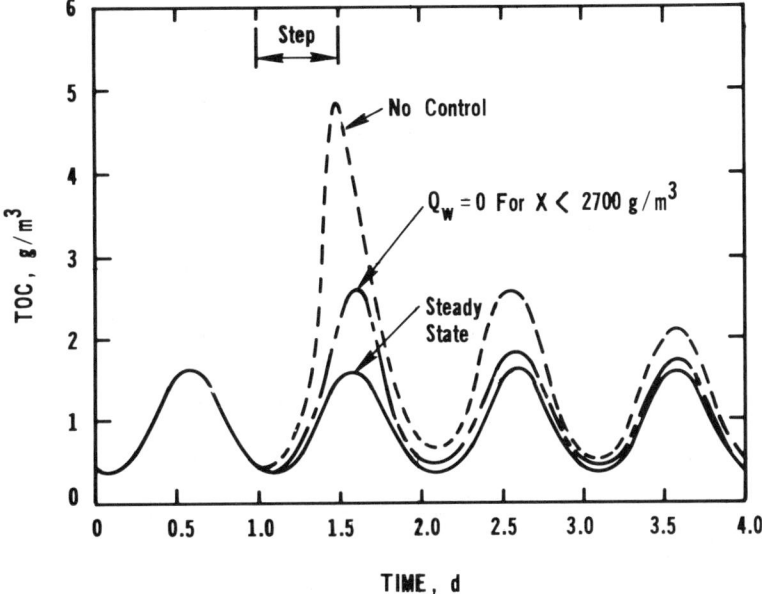

Figure 4. Soluble TOC response of CFSTR system to step change in flow with no change in mass organic input.

were step aeration, two were contact stabilization and one was an extended aeration system. Variables available for evaluation included flowrate, recycle rate, wastage rate, influent and effluent BOD and SS, MLSS, MCRT, F/M, sludge volume index (SVI), recycle sludge concentration and influent waste temperature.

The effect of variable and parameter variation on effluent BOD and SS concentrations was estimated using bivariate correlation analysis and multiple regression analysis. The Pearson correlation coefficient, r, and coefficient of determination, r^2, were used in the correlation analysis, and the multiple coefficient of determination, r^2, was used in the regression analysis. Typical results are shown in Tables I and II.

Note that the correlations are weak in all cases and that some of the results are counterintuitive. For example, a negative correlation between MCRT and effluent BOD and SS would be expected based on Equations 1 and 5, but a positive correlation exists for a number of plants. In general there is not a consistent pattern of correlation.

Results of the multiple regression analysis were similar to those for correlation analysis, as can be seen in Tables III and IV. In multiple regression analysis the coefficient of determination can be interpreted as

12 ACTIVATED SLUDGE PROCESS CONTROL

Table I. Correlation Analysis: Pearson Correlation Coefficient, r, Between Effluent BOD and Selected Variables

Plant number	SS_i	Flow	BOD_i	WTEMP	MCRT	F/M	MLSS	X_R	Q_R	Q_W	DO	SVI	Time
1	0.54	0.15	0.03	−0.57	−0.01	0.01	0.03	0.01	0.19	−0.08	0.24	0.03	−0.47
2	0.51	0.06	0.14	0.10	0.29	−0.14	0.37	0.03	0.53	—	0.08	0.52	−0.18
3	0.49	−0.07	0.20	−0.37	−0.03	0.25	−0.21	−0.10	—	−0.16	−0.21	−0.02	−0.27
4	0.82	−0.18	0.34	0.39	−0.46	0.36	−0.14	−0.21	−0.20	−0.21	—	0.34	0.36
5	0.62	−0.24	0.55	−0.09	—	—	0.09	−0.14	0.25	—	−0.31	0.31	0.39
6	0.60	−0.28	0.49	0.04	—	—	0.08	0.01	0.10	—	0.28	0.33	0.37
7	0.67	−0.07	0.40	−0.15	—	—	0.08	0.07	0.14	—	0.10	0.21	0.16
8	0.34	0.26	0.26	—	—	0.35	0.19	—	0.12	—	0.17	0.16	0.25
9	0.26	0.42	0.48	−0.55	−0.23	—	0.32	0.43	−0.17	—	−0.27	−0.33	−0.19
10	0.65	−0.60	−0.27	−0.62	0.38	—	0.51	0.53	−0.25	—	0.21	−0.28	−0.63
11	0.53	0.33	0.16	−0.07	−0.12	—	−0.07	0.01	0.25	—	—	−0.29	−0.06
12	0.43	−0.11	−0.03	−0.34	−0.01	—	0.16	−0.05	0.29	—	−0.37	−0.19	−0.54
13	0.36	−0.04	0.05	−0.02	—	—	0.07	—	—	0.05	—	−0.07	−0.05
14	0.28	−0.08	0.41	0.27	−0.08	—	0.11	0.02	0.02	0.09	−0.13	−0.22	0.28
15	−0.03	0.02	−0.06	0.12	−0.08	—	−0.09	—	—	—	—	—	0.13
16	0.93	0.19	0.13	−0.40	−0.14	−0.01	0.38	0.28	0.10	−0.10	—	0.24	−0.37
17	0.41	0.01	0.34	—	−0.19	0.23	0.02	0.07	—	−0.02	—	0.02	−0.10
18	0.58	−0.09	0.36	−0.15	0.01	0.01	0.24	0.25	0.14	−0.05	—	0.11	−0.45
19	0.35	−0.09	0.33	−0.18	−0.14	0.31	−0.13	−0.14	−0.04	0.20	—	0.10	−0.19
20	0.84	0.06	0.25	−0.36	—	0.09	0.36	0.26	0.27	0.10	—	0.10	−0.29
21	0.95	0.17	0.04	−0.26	—	0.03	0.21	0.09	0.34	0.04	—	0.05	−0.27

Table II. Correlation Analysis: Pearson Correlation Coefficent, r, Between Effluent SS and Selected Variables

Plant number	Flow	SS$_i$	WTEMP	MCRT	F/M	MLSS	X$_R$	Q$_R$	Q$_W$	DO	SVI	Time
1	0.14	0.18	-0.19	-0.11	-0.01	0.51	0.12	0.14	-0.08	0.08	-0.12	-0.11
2	-0.04	-0.03	-0.17	0.18	-0.20	0.32	0.03	0.39	—	0.09	0.42	-0.16
3	-0.12	-0.03	-0.51	-0.15	0.03	-0.30	-0.12	—	0.10	-0.08	-0.33	-0.25
4	-0.20	0.25	0.15	-0.58	0.28	-0.12	-0.18	0.14	-0.12	—	0.38	0.41
5	0.12	0.06	-0.07	—	—	-0.02	-0.09	0.05	—	-0.17	0.21	-0.01
6	0.10	0.06	-0.07	—	—	-0.19	0.06	0.07	—	-0.12	0.18	-0.07
7	0.14	0.15	-0.23	—	—	0.02	-0.10	0.21	—	0.10	0.18	-0.03
8	0.22	0.11	—	—	0.17	0.07	—	0.06	—	0.06	0.20	0.17
9	-0.31	0.23	-0.12	0.01	—	-0.24	0.10	0.34	—	0.03	-0.57	0.50
10	-0.53	-0.08	-0.37	0.14	—	0.32	0.37	0.13	—	0.38	-0.21	-0.46
11	0.41	0.04	-0.52	-0.23	—	0.03	0.23	0.12	—	—	-0.43	-0.41
12	0.01	-0.04	0.03	0.09	—	-0.08	-0.15	0.14	—	0.04	-0.12	-0.03
13	0.12	0.19	-0.20	—	—	0.05	—	—	-0.11	—	0.15	-0.16
14	0.19	0.18	-0.15	-0.36	—	0.10	0.21	0.04	-0.05	-0.21	-0.06	0.03
15	-0.13	0.09	-0.16	0.07	—	0.11	—	—	—	—	—	-0.23
16	0.22	0.11	-0.36	-0.15	-0.03	0.39	0.29	0.15	-0.12	—	0.18	-0.36
17	0.07	0.06	—	-0.41	0.17	0.16	-0.19	—	0.06	—	0.16	0.12
18	-0.27	0.30	-0.22	0.05	-0.08	-0.03	0.08	0.19	-0.24	—	-0.08	-0.40
19	0.24	0.05	-0.34	-0.08	0.27	-0.14	-0.10	0.23	0.13	—	0.26	-0.30
20	0.11	0.13	-0.22	—	0.04	0.18	0.05	0.37	0.11	—	0.15	-0.13
21	0.21	-0.04	-0.23	—	-0.02	0.16	0.06	0.39	0.13	—	0.05	-0.22

Table III. Regression Analysis: Coefficient of Determination, R^2, for Effluent BOD and Selected Groups of Variables[a]

Plant number	Input[b]	WTEMP	Biological & Operational[c]	SVI	Time	All Variables[d]
1	0.03	0.32	0.09 (Q_R)	0	0.22	0.0.48 (WTEMP)
2	0.03	0.01	0.29	0.27	0.03	0.42 (SVI, MLSS)
3	0.05	0.14	—	0	0.07	0.21 (WTEMP)
4	0.16 (BOD_i)	0.15	0.25 (MCRT)	0.12	0.13	0.43 (MCRT, WTEMP, BOD_i)
5	0.30 (BOD_i)	0.01	—	0.10	0.15	—
6	0.25 (BOD_i)	0	—	0.11	0.13	—
7	0.17 (BOD_i)	0.02	—	0.04	0.02	—
8	0.11 (BOD_i)	0	—	0.03	0.06	—
9	0.34 (BOD_i, flow)	0.30	0.21 (X_R)	0.11	0.04	0.40 (WTEMP)
10	0.37 (flow)	0.38	—	0.08	0.40	0.61 (WTEMP, time)
11	0.12 (flow)	0.01	—	0.08	0	0.27 (flow time)
12	0.01	0.12	—	0.04	0.29	0.40 (time)
13	0	0	—	0	0	—
14	0.19 (BOD_i)	0.07	0.02	0.05	0.08	0.32 (BOD_i, SVI)
15	0	0.01	—	—	0.02	—
16	0.07	0.16	0.16 (MLSS)	0.06	0.14	0.32 (WTEMP)
17	0.12 (BOD_i)	—	—	0	0.01	—
18	0.14 (BOD_i)	0.02	—	0.01	0.20	0.34 (BOD_i, time)
19	0.10 (BOD_i)	0.03	0.05	0.01	0.04	0.13 (BOD_i)
20	0.07 (BOD_i)	0.13	—	0.01	0.08	—
21	0.03	0.07	—	0	0.07	—

[a]Values in parentheses are the most important variable (greater than 5% significance) in explaining the total variation.
[b]Input loads: flow, BOD_i.
[c]Biological and operational variables: MCRT, MLSS, X_R, Q_R.
[d]All variables: flow, BOD_i, WTEMP, MCRT, MLSS, SVI, time.

Table IV. Regression Analysis: Coefficent of Determination, R^2, for Effluent SS and Selected Groups of Variables[a]

Plant number	Input[b]	WTEMP	Biological & operational[c]	SVI	Time	All variables[d]
1	0.05	0.04	0.11 (Q_R)	0.01	0.01	0.21 (flow, SS_i)
2	0	0.03	0.17 (Q_R)	0.18	0.03	0.22 (SVI, MLSS)
3	0.01	0.26	—	0.11	0.06	0.41 (WTEMP, SVI)
4	0.10 (SS_i)	0.12	0.36 (MCRT)	0.14	0.17	0.52 (MCRT, SVI)
5	0.02	0.01	—	0.04	0	
6	0.01	0	—	0.03	0	
7	0.05	0.05	—	0.03	0	
8	0.05 (flow)	—	—	0.04	0.03	0.52 (SVI, time)
9	0.13 (flow)	0.01	0.16 (Q_R)	0.32	0.25	0.33 (flow)
10	0.29 (flow)	0.14	—	0.04	0.21	0.44 (WTEMP, Flow)
11	0.18 (flow)	0.27	—	0.18	0.17	0.05
12	0	0	—	0.01	0	
13	0.07 (SS_i)	0.04	—	0.02	0.03	0.26 (MCRT, flow)
14	0.11	0.02	0.20 (MCRT, X_R)	0	0	
15	0.03	0.03	—	—	0.05	0.31 (MLSS, time, MCRT)
16	0.06 (flow)	0.13	0.17 (MLSS)	0.03	0.13	
17	0.01	—	—	0.02	0.01	
18	0.12	0.05	—	0	0.16	0.36 (time, flow)
19	0.07 (flow)	0.12	0.11 (Q_R, MLSS)	0.07	0.10	0.22 (WTEMP, flow)
20	0.03	0.05	—	0.02	0.02	
21	0.05	0.05	—	0	0.05	

[a]Values in parentheses are the most important variable (greater than 5% significance) in explaining the total variation.
[b]Input loads: flow, SS_i.
[c]Biological and operational variables MCRT, MLSS, X_R, Q_R.
[d]All variables: flow, SS_i, WTEMP, MCRT, MLSS, SVI, time.

the fraction of observed variation explained by a variable or group of variables. The values are additive and if all pertinent variables are included they will sum to 1.0. In most cases wastewater temperature was the most important variable. Controllable parameters (recycle rate, MCRT, MLSS) were significant factors in explaining the effluent BOD variation in 5 of the 21 plants, and in explaining the SS variation in 7 of the 21 plants. Input variables (BOD_i, SS_i and Q) were significant factors in 13 of the plants with respect to effluent BOD variation and 8 of the plants with respect to SS variation.

The conclusion that application of conventional theory cannot explain the variability in activated sludge process is inescapable. Why this is so is explainable in part by considering the nature of the models and the charateristics of the data.

WHY MODELS AND FACT DISAGREE

The disagreement between models and actual performance is, to a degree, an incorrect assumption. There is no model for predicting effluent suspended solids concentration, only an expectation that by using customary design values that the effluent SS will be less than 20 g/m^3. Effluent SS are the major contributors to effluent BOD and because this value is reported as total or combined rather than filtered and unfiltered the relationship between process parameters and effluent BOD is clouded. For example, an effluent BOD might be 20 g/m^3 of which 5 g/m^3 passes a 1-um filter. Increasing the MCRT value could lower the filtrate BOD significantly (1 g/m^3 is 20%) without causing a measurable change in total BOD. Thus there is much open to question about the available information on system response. It should be remembered that behavior of controlled laboratory systems is reasonably well described by the models (with the exception of SS), and thus it is unreasonable to throw them out altogether.

Causes of Variation in the Real World

Variation in effluent quality can result from a number of sources. Effects of loading, temperature, MCRT, recycle and chemical addition are recognizable if not always quantifiable. Factors such as wind on currents in sedimentation tanks, thermal stratification of sedimentation tanks, cyclic progression of microbial populations on a short-term or seasonal basis are much more difficult to deal with, however.

Instability of treatment systems is another possibility. There is an inherent design assumption that activated sludge processes always move toward a stable steady state. This is an assumption that is not well verified, and it is not unreasonable to conclude that a cyclic instability is the natural condition of activated suldge processes, perhaps because they go through microbial species progression. Some of these characteristics would be mitigated by normal operating procedures, but the philosophy of control is different if a system is viewed as inherently unstable as opposed to stable.

Time Scale Response

Most plant-scale data are from daily composite samples, whereas the transient response models are based on instantaneous conditions. The response time predicted by the models is of the order of minutes, as can be seen in Figures 2 and 3. Reaction- or kinetic-based models are suitable for predicting response to input variations, and if integrated and time averaged the results should be similar to prototype responses for systems that are not controlled by other factors such as changes in population and instabilities in the sedimentation tank.

Biological changes occur on a time scale of days. Pinpoint floc or filamentous cultures do not appear overnight. Operators' observations as well as SVI values usually show changes occuring over periods of a week or more. Although these changes are not currently predictable, physical observation allows the operator to take countermeasures. Both time scale and the fact that countermeasures are taken destroy any opportunity to model the process statistically because the events are (1) unrelated to known parameters and (2) nonrandom. Thus where biological changes are the dominant factor in determination of effluent quality, kinetically based control strategies are not going to have a good correlation. The best strategies are then those designed for acute response, chemical addition and change in process configuration being the most common but not necessarily the most appropriate.

A METHOD OF APPROACH

Mathematical models of biological wastewater treatment processes are very useful for the determination of design values such as oxygen uptake rates, sludge production and filtered effluent BOD. These models, and the control parameters resulting from their use, are not particularly helpful in estimating actual effluent quality because that value is dominated by suspended solids.

Mitigation of short-term effluent quality variation through kinetically derived process control parameters is not appropriate because the parameters are based on long-term, steady-state concepts and because the response of the activated sludge process is highly damped, as shown in Figures 2 and 3. Thus the question of the usefulness of on-line, process control strategies is not answered by current theory.

Two strategies are available for controlling effluent BOD and SS, but both are preventive and long-term in nature. The first involves matching process configuration with wastewater characteristics, and the second is an application of statistical tools for the determination of problems inherent in the system. The two strategies can both be used on the same system at the same time.

Conrol by Process Configuration

Producing low filtrate BOD values is not a problem in the activated sludge process, but consistently producing low total BOD and low SS values is a problem. Thus any process control strategy should focus on producing good solids separation. Four factors are of concern:

1. pinpoint floc
2. dispersed growth
3. filamentous bulking
4. hydraulic instabilities

The first three are kinetically related factors. Pinpoint floc occurs in cultures with very long SRT and filamentous growth occurs in carbon-limited systems, such as naturally occur in CFSTR activated sludge processes, at conventional SRT values. The best control procedure is to design systems to operate in the conventional SRT range and incorporate a non–carbon–limited region into the system. Ideal plug flow processes have such a system, but real processes often have excessive backmixing. This can be eliminated by baffling, or inserting a selector [29] as shown in Figure 5. If the non–carbon– limited region is also anoxic a further kinetic advantage is given to floc-forming organisms. It is worth mentioning that floc formation appears to be a function of long-term average conditions, i.e., SRT values.

Sedimentation tank instabilities are best prevented by baffling. Proper placement of baffles mitigates density stratification and current instability, resulting in behavior nearer to the ideal.

Use of Statistical Tools in Control

Niku and Schroeder [30] have proposed the use of a statistical parameter, the coefficient of reliability (COR) and a definition of stability

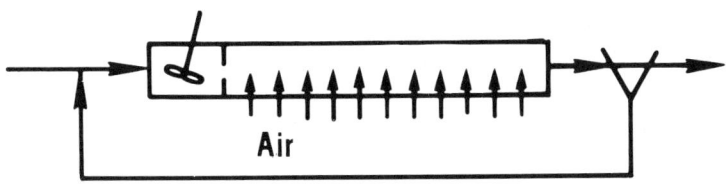

a) Activated Sludge System With Anoxic Selector Section

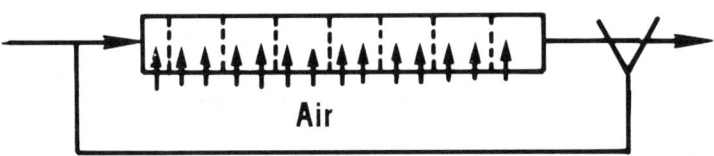

b) Activated Sludge System With Baffleing to Decrease Longitudinal Dispersion

Figure 5. Suggested configurations for control of bulking

based on statistical concepts and experience as process control tools. The COR is the ratio of the annual mean BOD or SS value that must be met if a given plant is to meet an effluent standard a certain percentage of the time, to the standard value:

$$COR = m_x/S \qquad (6)$$

where m_x = annual mean variable value
       S = required standard value

For example, if a plant were required to meet a 30 g/m³ effluent BOD value 99% of the time, the annual mean value would need to be considerably lower than 30 g/m³. The value would be found from a statistical evaluation of the plant's historical performance if the plant were in operation, and from generalized activated sludge plant data if the plant was under design.

Equation 6 can be stated in any way desired. Current standards are based on 30- and 45-day averages and the COR can be set up to fit either running or calendar values, both of which are less stringent than a reliability based on daily values.

Niku et al. [29] found in a study of daily composite data from 43 activated sludge plants that the log normal distribution provided the best fit, and they based their calculations of the COR on that distribuiton. They put the relationship in terms of the arithmetic coefficent of determination, i.e., the ratio of the sample standard deviation, s, to the sample mean, m_x:

$$V_x = s/m_x \qquad (7)$$

A graph of the relationship is presented in Figure 6. Note that the abscissa can be interpreted as the normalized mean or the COR.

The stability parameter was proposed by Niku and Schroeder [30]. As stated above it is based on experience—in this case the examination of data from the same 43 plants. A standard deviation of 10 g/m³ of either BOD or SS was chosen as a stability cutoff on the basis of the mean and range of the annual data. Plants having a variable standard deviation greater than 10 g/m³ are considered unstable.

Application of the work of Niku and co-workers to process control requires that a data record for the particular plant be constructed. As many parameters and variables as possible are included in the record, but the primary ones are effluent BOD and SS. Reliability and stability are then estimated using the available data. Seasonal or shorter cycle trends are looked for using nonlinear fits or time series analysis. A conclusion is then made as to the acceptability of plant performance. If the performance is unacceptable, an attempt to determine if variability in process parameters or variables can explain the results. In many cases a control parameter such as overflow rate can be related to performance, and this method of control will prove useful. At the very least the actual performance of the plant can be established.

A continuing record of performance is useful, also. Trends in reliability and stability can suggest that upset prevention measures need to be taken. If the 30-day running standard deviation begins to increase, instability problems may be beginning and a procedure such as stopping wastage, increased recycle or recycle chlorination might be appropriate in a particular plant.

CONCLUSIONS

1. Conventional control parameters do not correlate with effluent quality.
2. Conventional control parameters are unrelated to the principal causes of effluent quality variation.

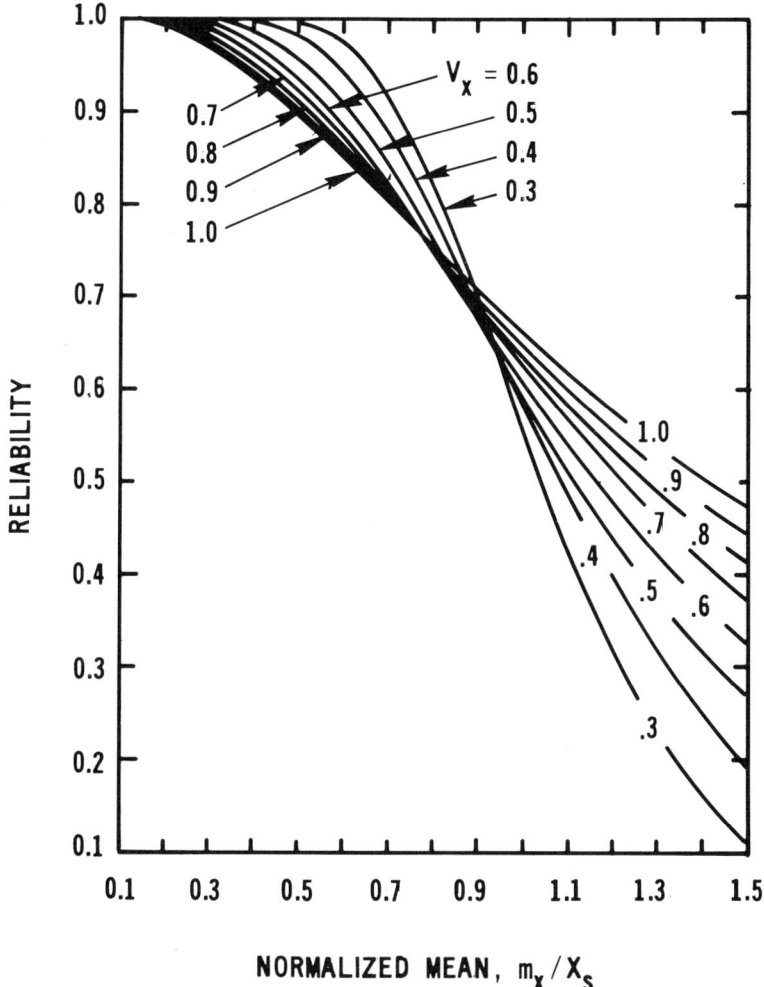

Figure 6. Reliability vs normalized mean for different coefficients of variations.

3. Variation in effluent quality resulting from short-term variations in influent characteristics is not a major problem.

4. The principal control problem in activated sludge processes, with respect to effluent quality, is the loss of suspended solids in the effluent.

5. Preventive control measures that enhance the competitive advantages of floc-forming bacteria while maintaining the culture in a dominant floc-forming growth phase are the best means of control.

6. Statistical analysis of the data records for individual plants provide a useful way to evaluate performance.

For further information on this topic, see the Questions and Answers section on page 211.

REFERENCES

1. Metcalf and Eddy, Inc. *Wastewater Engineering*, 2nd ed., (New York: McGraw Hill Book Company, 1979).
2. Schroeder, E.D. *Water and Wastewater Treatment*, (New York: McGraw Hill Book Company, 1977).
3. Sundstrom, D.W., and H.E. Klei. *Wastewater Treatment* (Englewood Cliffs, N.J.: Prentice-Hall, Inc., 1979).
4. Eckenfelder, W.W., and D.J. O'Connor. *Biological Waste Treatment*, (New York: Pergamon Press, 1961).
5. McKinney, R.E. *Microbiology for Sanitary Engineers*, (New York: McGraw Hill Book Company, 1962).
6. Downing, A.L., A.G. Boon and R.W. Bayley. "Aeration and Biological Oxidation in the Activated Sludge Process," The Institute of Sewage Purification, Brighta Conference (1961).
7. Gaudy, A.T., and B.G. Turner. "Effect of Air Flow Rate on Response of Activated Sludge to Quantitative Shock Loading," *Journal Water Pollution Control Federation*, 36:767 (1964).
8. Heukelekian, H., and R.S. Ingols. "Studies in Activated Sludge Bulking," *Sew Works J.* 12:694 (1940).
9. Sezgin, M., D. Jenkins and D.S. Parker. "A Unified Theory of Activated Sludge Bulking," *J. Water Poll. Control Fed.* 50:362 (1978).
10. Silverstein, J.A., and E.D. Schroeder. "Performance of SBR Activated Sludge Processes with Nitrification/Denitrification," submitted to *J. Water Poll. Control Fed.* December 1981.
11. Chambers, B. "Effect of Longitudinal Mixing and Anoxic Zones on Settleability of Activated Sludge" in Chambers, B. and E.J. Tomlinson, Ed. *Bulking of Activated Sludge: Preventative and Remedial Methods* (Chichester, England: Ellis Horwood Ltd, 1982).
12. Eikelboom, D.H. "Biosorption and Prevention of Bulking by Means of High Floc Loading in Chambers, B. and E.J. Tomlinson, Ed. *Bulking of Activated Sludge* (Chichester, England: Ellis Horwod Ltd., 1982).
13. Van Den Eynde, E., J. Houtmeyers and H. Verachtert. "Relation Between Substrate Feeding Pattern and Development of Filamentous Bacteria in Activated Sludge," in Chambers, B. and E.J. Tomlinson Ed. *Bulking of Activated Sludge*, (Chichester, England: Ellis Horwold Ltd, 1982).
14. Eckhoff, D.W., and D.I. Jenkins. "Transient Loading Effects in the Activated Sludge Process," Third Int. Conference on Water Pollution Research, Munich, (1966).
15. Selna, M.W., and E.D. Schroeder. "Response of Activated Sludge Processes to Organic Transients," *J. Water Poll. Control Fed.* 50:944 (1976).
16. Storer, F.F., and A.F. Gaudy. "Computational Analysis of Transient Response to Quantitative Shock Loadings of Heterogeneous Populations in Continuous Cultures," *Environmental Science and Technology* 3:143 (1969).

17. Dennis, R.W., and R.L. Irvine, "Laboratory Investigation of Fill and Draw Biological Reactors for Treatment of High Strength Wastes," *Proceedings of the 32nd Industrial Waste Conference, Purdue University* (Ann Arbor, MI: Ann Arbor Science Publishers, Inc., 1978).
18. Irvine, R.L. "Sequencing Batch Reators", presented at the U.S. EPA Seminar on Emerging Systems, San Francisco, April 1982.
19. Irvine, R.L., and A.W. Busch. "Sequencing Batch Reactors," *J. Water Poll. Control Fed.* 51:238 (1979).
20. Silverstein, J.A. "Operational Control of Bioflocculation in a Batch Wastewater Treatment System," Doctoral Dissertation, Department of Civil Engineering, University of California, Davis (1982).
21. McLellan, J.C., and A.W. Busch. "Hydraulic and Process Aspects of Reactor Design—II—Response to Variations in Loading," *Proceedings of the 24th Industrial Waste Conference, Purdue University* (West Lafayette, IN: Purdue University, 1970).
22. Schroeder, E.D., and J.A. Silverstein. "Interactive Models as Tools in Control of Activated Sludge Systems," presented at International Confernece on Fundamental Aspects of Wastewater Treatment, Graz, Austria, October, 1980.
23. Adams, B.J., and R.S. Gemel. "Performance of Regionally Related Wastewater Plants," *J. Water Poll. Control Fed.* 45:2008 (1973).
24. Hann, R.W., T.W. Sparr, H.W. Wolf and D.J. Schaezler. "Evaluation of Factors Affecting Discharge Quality Variation," report to Texas Water Quality Board, 1972.
25. Carr, D.F., and J. Ganczarczgk. "A Performance Analysis of An Activated Sludge Treatment Plant," Department of Civil Engineering, University of Toronto (1971).
26. Thoman, R.V. "Variability of Waste Treatment Plant Performance" *J. San. Eng. Div.* ASCE 96:SA3 (1970).
27. Townshend, A.R. "Statistical Analysis of the Effluent Quality of Biological Sewage Treatment Processes," *Proceedings of the 3rd Canadian Symposium on Water Pollution Research*, Toronto (1968).
28. Hovey, W.H., E.D. Schroeder and G. Tchobanoglous. "Optional Size of Regional Wastewater Treatment Plants," California Water Resources Center Contribution No. 161, 1977.
29. Niku, S., E.D. Schroeder and F.J. Samaniego. "Performance of Activated Sludge Processes and Reliability Based Design," *J. Water Poll. Control Fed.* 51:2841 (1979).
30. Niku, S., and E.D. Schroeder. "Factors Affecting the Effluent Variability From Activated Sludge Processes," *J. Water Poll. Control Fed.* 53:546 (1981).
31. Chudoba, J., V. Ottova and V. Madera. "Control of Activated Sludge Filamentous Bulking—I." *Water Res.* 7:1163 (1973).

CHAPTER 2

PROCESS CONTROL STATEGIES

Dr. Robert M. Arthur
    Arthur Technology, Inc.
    Fond du Lac, Wisconsin

INTRODUCTION

The activated sludge process is the most widely used biological wasterwater treatment method. Since the beginning of the 20th century, when the process was developed in a laboratory of a treatment plant in Manchester, England, the practice of treating wastewater by the activated sludge process spread rapidly throughout the world. In the past seventy years the original concept has undergone many modifications, but throughout all these changes the process still depends on the coupling of the biological process of metabolism and the physical process of solids separation. Although much has been done to improve the equipment used in the treatment process, only cursory attention has been paid to *practical* methods of controlling the process. This is not to say that studies to determine the kinetics of the treatment process have not been made, for this is a fertile field for many professors and graduate students. The transfer of their findings to everyday plant operation, however, has not been successful because the assumptions made in the laboratory hardly ever agree with what actually happens in an operating plant.

The fact that it has not been possible to put theory into practice does not eliminate the need to find better means of controlling this most widely used process of treating wastewater [1]. Recent studies by EPA [2] have found that failures in plant operation are in many cases due to lack of knowledge about how to apply test data to control the process. The utilization of test data in a logical sequence of reactions to modify a process and maintain a specified result is a process control strategy—the subject of this chapter.

In discussing process control strategies it is reasonable to ask three questions:

1. What process is to be controlled?
2. What in the process can be controlled?
3. What are the control strategies?

THE PROCESS TO BE CONTROLLED

The activated sludge process is a complicated combination of two of nature's most basic phenomena—the natural breakdown of organic matter by biological metabolism and the separation of solids and liquids by the natural force of gravity. Although these two phenomena have been around much longer than the activated sludge process, it is strange that so little of what we know about these phenomena is actually applied to improving the operation of wastewater treatment plants.

Biological Phenomena

The biological metabolism of foods is fundamental to sustain any living cell. As shown in Figure 1, food is the source of energy for the cell and building material for growth of new cells. And it makes no difference if the cell is in the tissue of animals or freely suspended in activated sludge—the basic phenomena is the same. Although it may appear to be a very simple process, it is indeed a complicated series of biochemical reactions designed to produce energy. The energy produced may in turn be used to generate heat, provide mobility, and assemble new cell structure. It is not a physical oxidation (burning process) but a highly regulated step-by-step process which extracts energy in the form of high-energy hydrogen atoms (H^+). The high energy is transferred from hydrogen to adenosine triphosphate (ATP) which then transports the energy to a place of energy use, as shown in Figure 2. Note that the process may be anaerobic or aerobic depending on one factor—the amount of dissolved oxygen (DO) present. If the DO is low, the process generates organic alcohols and acids as products. If the DO is plentiful, the process generates only water. Note the carbon dioxide (CO_2 plays no important role and is released as a by-product in the process.

The important points to remember are:

PROCESS CONTROL STRATEGIES   27

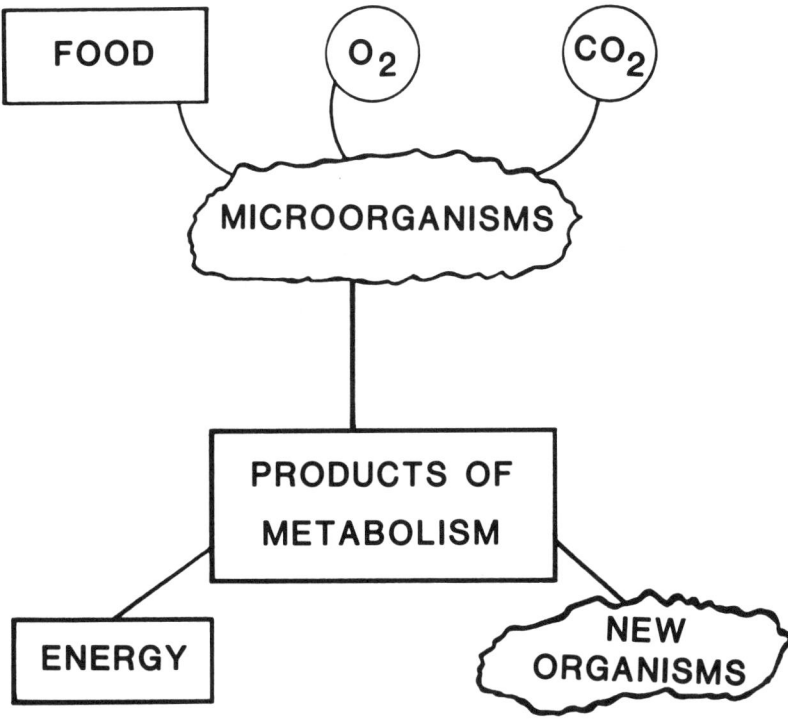

Figure 1.   Cellular metabolism.

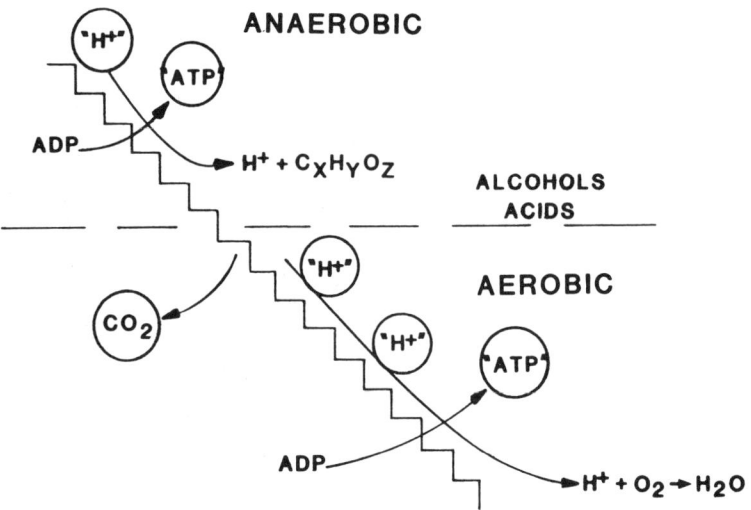

Figure 2.   Anaerobic-aerobic energy generation

1. The process is a regulated system for producing energy.
2. In many cells the system will operate as either an aerobic or anaerobic process.
3. Energy is extracted as high-energy hydrogen atoms which transfer the energy to molecules of ATP.
4. In an aerobic process the final acceptor of hydrogen atoms is oxygen.

With the above facts in mind, it is appropriate to recognize that the cell is also capable of some degree of control of the metabolic process. The primary stimulus to initiate the process is the presence of food. As shown in Figure 3, food triggers the release of extra cellular enzymes—biological catalysts that break down the food, by digestion, into fragments of glucose and amino acids which pass through the cell wall to enter the metabolic system. To complete aerobic metabolism it is necessary for dissolved oxygen to be present and for CO_2 to be released. The complete process is therfore regulated by a number of actions, including:

- presence of food
- discharge of enzymes
- enzymatic breakdown of food into fragments
- transfer of fragments through the cell wall
- transfer of CO_2 out of the cell
- transfer of O_2 into the cell

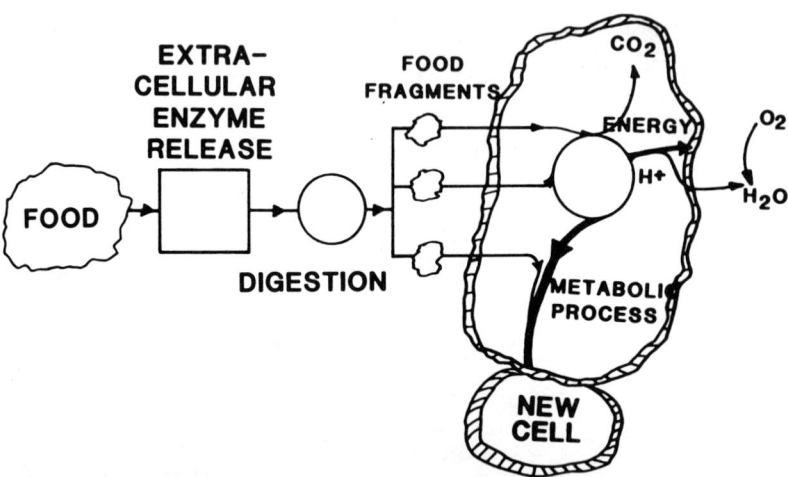

Figure 3. Digestion-metabolism of food.

It is clear that, assuming a well mixed healthy activated sludge, only two of the above actions really relate to controlling the activity of the process—the presence of food and oxygen. Of these, food is most important to continuing living processes since if oxygen is not present the cell could switch to an anaerobic process. *In spite of its importance, little attention has been paid to developing on-line techniques of monitoring food.*

Physical Phenomena

Nature's simple process of settling solids by gravity is an equally important part of the activated sludge process. The generation of a large mass of microorganisms by converting food to energy and new cell structure is a fine method of removing food from the wastewater, but it creates the problem of removing clean water from the biomass. Certainly the least expensive way to do this is to use nature's force of gravity and remove solids by sedimentation.

Sedimentation is nature's method of removing solids from gases and liquids. The physics of sedimentation have been thoroughly investigated and it is possible under certain conditions to theoretically determine the rate of sedimentation from the laws of mechanics. In general the rate of sedimentation is related to the weight of the solid F_W, bouyancy F_B, and the drag force F_D on the particle, as shown in Figure 4. From the laws of mechanics it is possible to derive the following formula for the velocity of settling:

$$v = \frac{1}{18} \frac{g}{\eta} (\rho - \rho) d^2$$

where
- v = settling velocity
- g = acceleration of gravity
- η = absolute viscosity of the fluid
- ρ = density of fluid
- ρ_1 = density of the particle
- d = diameter of the particle

Unfortunately the above formula only applies to well defined shapes having specific surface characteristics. For less well defined shapes it is necessary to include multipliers and other factors which are obtained from experiments. For irregular particles, such as found in activated sludge, the problem is very complex and the only way to determine settling rates is to perform a settling test.

In one of the earliest applications of activated sludge, the treatment tanks were batch fed and were used as both aeration and settling tanks.

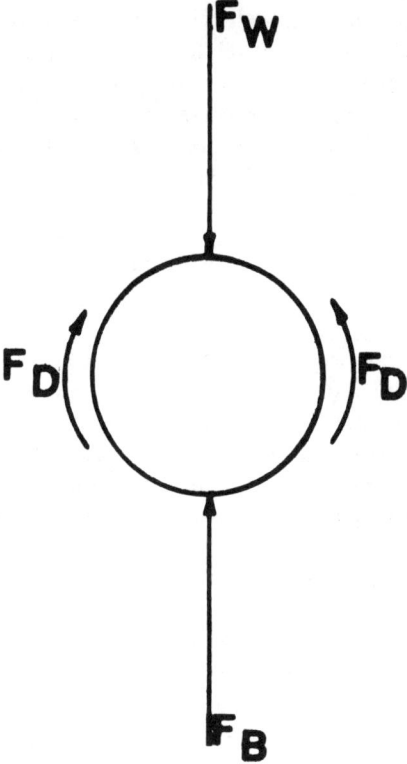

Figure 4. Forces of sedimentation

The tanks were first aerated and then the air was turned off and the solids were allowed to settle. This meant that only one force, the force of gravity, acted on the activated sludge particles (Figure 5). The only variable to be concerned about is the rate of settling.

Today, flow-through sedimentation tanks are used; this adds a complicating factor to the settling process. In addition to the force of gravity, solids are also subjected to a momentum force component created by the velocity of flow through the tank. As shown in Figure 6, this momentum component forces the particles along the tank at the same time gravity is forcing the particle to settle. The result is a particle path which is the resultant of the two forces. In effect, the magnitude of each force determines the direction and rate of settling in the tank.

Unfortunately, the magnitude and direction of the momentum force are impossible to calculate theoretically and difficult to determine experimen-

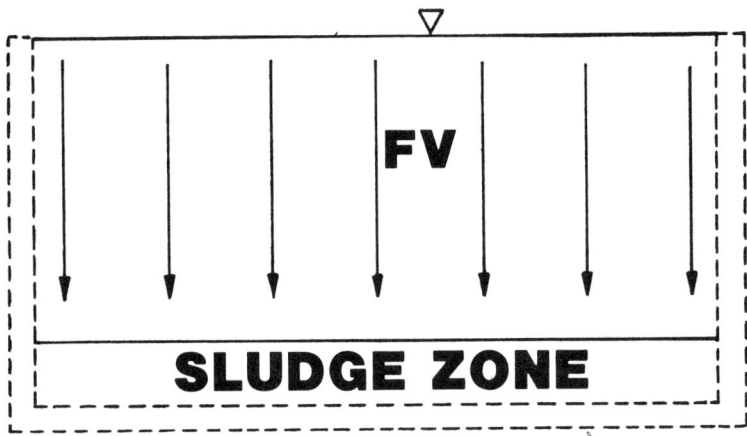

Figure 5. Batch settling tank.

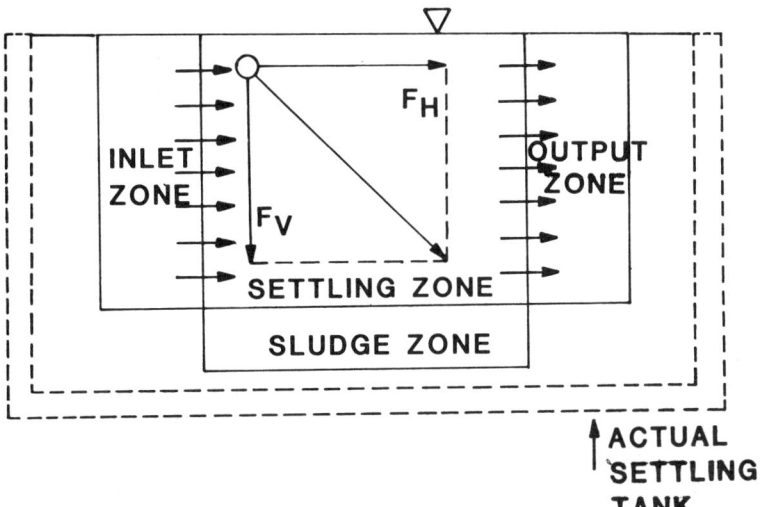

Figure 6. Flow-through settling tank.

tally. This means that there is really *no* way to determine the direction or rate of settling. This problem is compounded in a flow-through activated sludge settling tank because of the variations in flow which occur hourly and variations in the floc size, shape, and desity which may occur daily if not more often.

From the above it is clear that the physical process of settling in a flow-through tank, although not complicated, is almost impossible to analyze on a practical basis. In addition, the settling test, which is about the only technique available to analyze the settling process, is not performed frequently enough to be of value in process control.

It is interesting to note that little *on-line* information has been used to measure the performance of the two phenomena which are fundamental to the activated sludge process, i.e., biological metabolism and gravity sedimentation. It seems unrealistic to expect that it is possible to control a process unless continuous *timely* information is available about changes in the process.

WHAT CAN BE CONTROLLED?

In a conventional activated sludge plant only three variables can be controlled. These are: rate of aeration, rate of return sludge and rate of wasting, as shown in Figure 7. These are also the only controllable variables in other types of activated sludge processes, including contact stabilization, extended air and pure oxygen. Although these variables are normally conisidered to be ontrollable, many plants do not have sufficient range of control of these variables or a satisfactory scheme of control. In this case process *control* is virtually impossible.

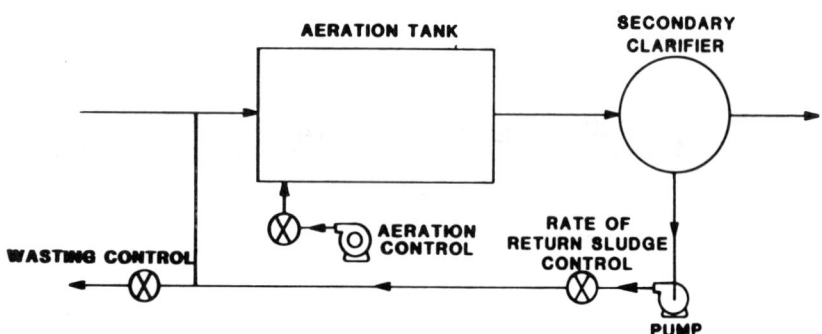

Figure 7. Controllable variables.

Control of Aeration

The ability to control the rate of aeration should be a primary consideration in plant design. This is especially true when it is recognized that the energy requirements of aeration equal 60 to 80% of all energy costs to operate the plant, and that with control, savings of from 25 to 50% are possible. Figure 8 is a graph of changes in respiration obtained by sampling with an on-line respirometer at the inlet of the aeration tank of an activated sludge wastewater treatment plant. Note the wide variations in demand for oxygen each day and throughout the week. The solid line at the top of the graph indicates the amount of oxygen supply that would be needed if there were no control of aeration. The crosshatched area represents the amount of oxygen (or energy) that is wasted.

Although it is clear that a considerable amount of energy can be saved with aeration control, it is strange that aeration control is not practiced in most plants. The reason is that few plants have the capability to control aeration. With mechanical aeration systems, variation in aeration can only be achieved with two-speed or variable-speed motors. And with diffused air systems, the capability to vary aeration is dependent on the type of blower and the type of diffuser. For example, a positive displacement blower operates most efficiently at one speed, so the best way to vary output is to have several blowers of different sizes and only utilize the blower or combination of blowers that satisfies a given oxygen demand. Also, it must be recognized that all diffusers cannot be used with varying air supply, i.e., fine bubble diffusers will clog at low air flows.

One type of aeration control which is available at plants having mechanical or coarse bubble diffusers but is *not used* is to turn the aerator off when the demand for oxygen is low. The limitation to the length and frequency of off-time is the respiration rate and the settling rate of the sludge. The off-time must first be controlled by relating the respiration rate to the amount of dissolved oxygen available. A second factor in control of off-time is sludge compaction; the activated sludge should not be allowed to settle to a state of compactness which will prevent the sludge from resuspending when the aerator is turned on.

Control of Return Sludge

In the original batch method of activated sludge there was no return sludge. After the aerators were turned off and the settling took place, the supernatant was discharged as final effluent and the settled sludge was disposed of by any convenient method. When it was recognized that the sludge contained a well developed and acclimated microorganism system which could be used again, return sludge became part of a conventional system and flow-through tanks became common.

If it is recognized that sludge is returned to reuse an existing culture of microorganisms, then it is logical to ask, "How much sludge should be returned?" The answer to this must be based on the basic biological concepts discussed above, which pointed out the importance of food in controlling the metabolic system. In an activated sludge process, the food

34 ACTIVATED SLUDGE PROCESS CONTROL

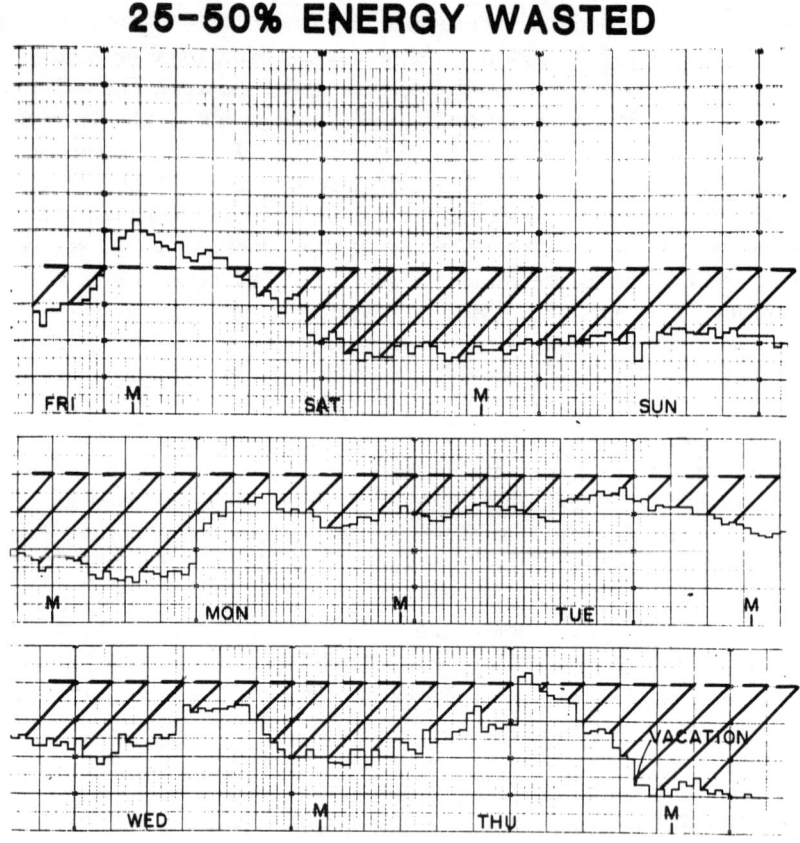

Figure 8. Energy wasted.

is the organic matter in the raw wastewater and for best operation the amount of food must be balanced by the proper concentration of viable microorganisms, i.e., the food to microorganism ratio (F/M). If F/M is too high, the microorganisms are saturated with food and the microorganisms are just not capable of eating all the food. If F/M is too low, all microorganisms in the sludge will not be fed resulting in a loss in viable cell mass.

Matching microorganisms with food by varying the rate of return sludge is not an easy task. One fact that must be known is the number of viable microorganisms per unit volume in the return sludge, which is a function of the settling characteristics of the sludge. A specific pumping rate in no way guarantees that a specific number of microorganisms will be present to match a specific amount of food. Another fact is that increasing return sludge rates may satisfy a requirement for more microorganisms in an initial period, but the effect is minimized in a short time due to a redistribution of solids concentration in all parts of the system.

Control of return sludge to match food with viable microorganisms is further complicated by the fact that the conventionl F/M ratio is not a valid control strategy. The use of BOD_5 to indicate food entering the aeration tank bears little relationship to the actual concentration and type of food entering the tank. A highly diluted environmentally controlled sample in a bottle cannot be expected to indicate what is taking place in a highly agitated tank with large concentrations of microorganisms. Of course, an obvious disadvantage in process control is the 5-day delay in getting the results. In addition, it is clear that the use of mixed liquor volatile suspended solids (MLVSS) is a poor indication of the concentration of *viable* microorganisms [4].

Substitution of rapid *physical* methods of determining food may work on *some* wastewaters but the measurements in no way represent quantities of food as "seen" by the microorganisms.

Just as with aeration control, the need for F/M control appears obvious but is almost impossible to achieve in existing activated sludge systems.

Control of Wasting

The basic biological system described above indicated that part of the food utilized by microorganisms was to build new cell structure. Synthesis of new cells is also a part of the activated sludge process. In the case of activated sludge the growth of new cells must be controlled because the metabolic requirements of a large cell mass may exceed the physical capabilities of the plant. In most activated sludge plants the limitations to increased cell mass are the capability to supply oxygen and the capability to effectively settle the biomass. In addition to limiting cell mass, wasting also accomplishes removal of cellular waste products. With these capabilities and limitations in mind, plants are normally designed to operate at a given concentration of biomass and therefore excess cell growth must be wasted.

In most modern day activated sludge plants wasting is accomplished by directing a portion of return sludge to a point of disposal. In many plants the operation of wasting is performed intermittently and in other plants only for a certain time period each day. The amount wasted is generally based on the magnitude of total biomass in the system. The practices of wasting intermittently and in accordance with biomass concentration are in opposition to the basic biology of the process.

Wasting intermittently drastically changes the fundamental relationship between food and quantity of microorganisms. When wasting is taking place, the quantity of microorganisms is reduced, placing a metabolic strain on those microorganisms still present in the system. The higher ratio of food overloads the metabolic system and creates an imbalance and a cellular growth phase. The elimination of wasting creates the opposite imbalance. The result of intermittent wasting is a continuous imbalance between food and microorganisms and a disturbed process.

Using the concentration of total biomass to determine the amount of wasting is not an acceptable plan either, for it bears little relationship to the viability of the process. Wasting is performed to eliminate excessive microorganisms that demand oxygen and to eliminate inorganic compounds or complicated organic compounds which cannot be metabolized or can only be partially metabolized. The use of total mass of solids does not give a good indication of what should be wasted and the effect of wasting on the process.

Additional Controllable Components

As shown in Figure 9, there are two additional components that could be used in an activated sludge plant to facilitate improved operation. These are [5]:

1. storage of raw wastewater
2. storage of microorganisms

Storage of raw wastewater has obvious advantages in the control of activated sludge processes. Storage should be designed to equalize both food and flow into the treatment process. If complete equalization is accomplished, the activated sludge process would operate in a steady-state mode, and rates of aeration, return sludge and wasting would be constants. Complete equalization is normally not possible, but partial equalization of both food and flow can contribute to better control of activated sludge.

Raw wastewater storage is also frequently used when pH adjustment and nutrient addition are necessary or when toxic materials are commonly discharged to the plant. For pH adjustment and nutrient addition,

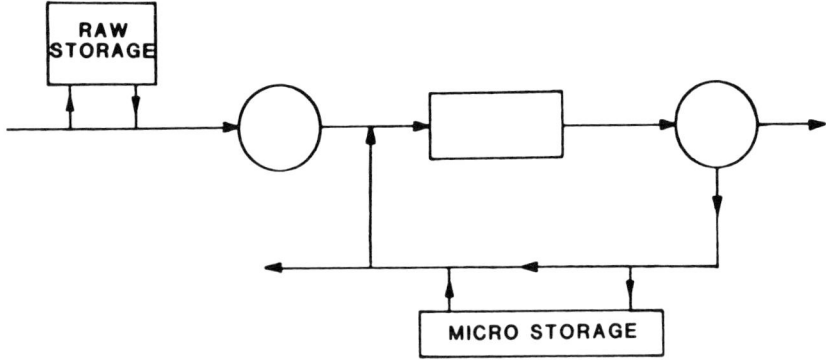

Figure 9. Additional controllable components.

on-line storage is common, but for toxic waste diversion, off-line storage is necessary. Since some "toxic" wastes can be treated in certain concentrations it may be possible to feed the stored wastes back into the plant if concentrations can be controlled.

Storage of microorganisms can be used for two purposes:

1. eliminate "washout" of solids during periods of high hydraulic loads caused by infiltration-inflow (I/I)
2. to stabilize the food to microorganism ratio

To eliminate "washout" of solids it is necessary to divert solids from the final tank and aeration tank when there is an increase in hydraulic load caused by I/I. When the excessive flow decreases, solids are immediately transferred back to the aeration tank. Aerated microorganism storage used for this purpose eliminates the need to generate new microorganisms each time there is a washout. It eiliminates the cycle of washout-reseeding-washout which occurs over several months in the spring of the year at many municipal plants.

Microorganism storage is the only truly effective method of stabilizing the food to microorganism ratio. To be effective the microorganisms must be thickened as well as stored. In use both food and microorganism concentration are measured using on-line respirometry. Changes in the concentration of food are used to control the flow of microorganisms from the microstorage facility.

CONTROL STRATEGIES

Definition

A control strategy as used in activated sludge means a method, technique or formula which is used to develop a response to an observation or series of observations. As with most physical or biological processes, adjustment of the controllable components of the process must be made in accordance with a plan or scheme, i.e., a strategy.

An activated sludge control strategy can range from the simple to the very complicated. It can also be manually or automatically implemented. The strategy could be initiated with a visual observation of grayish chunks of sludge rising in the final clarifier. The strategy is to increase the rate of return sludge by manually adjusting a variable-speed motor connected to the return sludge pumps. Or the strategy could be initiated by on-line sensors feeding signals to a microprocessor which manipulates and analyzes the data in accordance with a mathematical model and then automatically directs appropriate action in adjusting controllable components.

Examples of commonly used activated sludge control strategies include:

- sludge volume index (SVI)
- food to microorganism ratio (F/M)
- sludge retention time (SRT)

In addition, there are other, less well defined techniques of control which may not be classified as strategies but nevertheless are still excellent means of control. These include:

- sludge settling and rise rate
- microscopic examination of floc
- visual observation of sludge characteristics

In effect, anything that helps to direct a reaction to an action can be considered a control strategy.

Application of Control Strategies

Control strategies are developed to provide information which can be used to adjust controllable variables. As indicated above, an activated sludge process has three controllable variables:

1. rate of aeration
2. rate of return sludge
3. rate of wasting

Control strategies will be examined in relation to their application to the control of these variables.

Rate of Aeration

Controlling the rate of aeration has in the past been almost ignored by design engineers and therefore operators found no need to develop a control strategy. Today however, the increased cost of energy has forced designers and operators to take a new look at control of aeration.

The designer of course must weigh such factors as oxygen transfer efficiency, maintenance and cost when selecting aeration equipment. But of equal or greater importance is the need to design controllability into the aeration system. To do this he must, with any means at his disposal, determine the hourly and daily changes in oxygen requirements [6].

Two methods of controlling aeration are shown in Figure 10. One method of controlling aeration consists of a feedback system in which dissolved oxygen changes in the aeration tank are used to indicate changes in oxygen requirements. A second method consists of a feed forward or *anticipatory* system using respiration rates which indicate the actual demand for oxygen in cubic feet of air per gallon of mixed liquor [1].

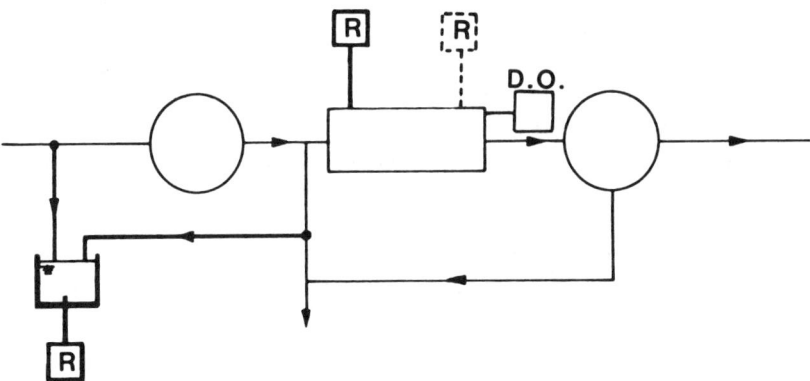

Figure 10. Control of aeration.

$$Q_{air} = \frac{1.114 \times 10^{-6}}{E} [RR]$$

where
- E = oxygen transfer efficiency in decimals
- RR = respiration rate of mixed liquor at inlet to aeration tank (ml/l/hr)
- Q_{air} = cubic feet of air per minute per gallon of mixed liquor under aeration.

Another reason for aeraion is to provide mixing of food, oxygen and microorganisms. As pointed out above, mixing is only required if food is present and there are high respiratory requirement. Turning the aerator off when these conditions do not exist should be a part of the total aeration control system.

Rate of Return Sludge

The traditional strategy to control return sludge is to use a fixed percentage of the plant flow. It is clear that the use of a percentage is an attempt to match the amount of food coming into the system with sufficient microorganisms. The logic is good but the method does not satisfy the objective, because changes in flow do not always represent changes in food or vice versa. This is almost always true in industrial plants and equally true in municipal plants having industrial contribution or infiltration-inflow. Using percentage of plant flow as a return sludge control strategy is an example of misuse of a physical measurement to indicate a biological characteristic.

A far better method of determining the rate of return is to look at what is supposed to happen biologically and physically in the system and then develop a control strategy to accomplish this objective. One objective for varying the return sludge is the traditional one—to stabilize the food to microorganism ratio. As pointed out previously, this may be an exercise in futility since increasing return sludge to supply additional microorganisms to balance an increase in food may provide a desired F/M for a limited time only.

The basic concept of food to microorganism ratio is still a valid and useful control strategy *if* F actually represents food as seen by the microorganisms, i.e., *biological food*, and M represents *viable* microorganisms. The best method is to let the microorganisms themselves indicate how much food is entering the aeration tank and the concentration of viable microorganisms in the return sludge. Measurements of respiration rates can be used to indicate both of these values.

The amount of food present is indicated by the respiration rate of the mixed liquor at the inlet to the aeration tank, and the concentration of viable microorganisms is indicated by the respiration rate of endogenously respiring return sludge. These respiration rates, when coupled with flowrates, can produce a new food to microorganism ratio:

$$\frac{F_b}{M_v} = \frac{RR_{ml_i}(Q + Q_{RAS}) - RR_{RAS} Q_{RAS}}{RR_{RAS} Q_{RAS}}$$

where
RR_{ml_i} = respiration rate of mixed liquor at inlet
RR_{RAS} = respiration rate of return sludge
Q = plant flow
Q_{RAS} = return sludge flow

(Note: RR and Q can be in any common units since they cancel out.)

Since Q_{RAS} is a variable in this formula it is possible to solve for this variable and thus have a new control strategy for return sludge based on information that can be obtained each half hour with on-line instrumentation.

$$Q_{RAS} = \frac{RR_{ml_i} Q}{F_b/M_v \, RR_{RAS} - RR_{ml_i} + RR_{RAS}}$$

Another objective for controlling the return might be to control the sludge blanket level in the final clarifier. Variation in sludge level is due to changing settling characteristics which in turn could be due to changes in biological conditions, i.e., changes in F/M or the presence of toxins or changes in physical conditions such as flow. In this case the control strategy is to maintain a given blanket level and this is accomplished by pacing return pumps with measurements of blanket level height.

A third objective may be to prevent anoxic or anaerobic conditions in the return sludge. This has important biological implications and can create unstable conditions in the bottom of the final tank. An anoxic condition can cause denitrification with the release of nitrogen gas and rising floc. Anaerobic conditions can also result in rising floc due to the generation of hydrogen sulfide.

To eliminate anoxic or anaerobic conditions it is necessary to control return sludge by matching the detention time of the sludge with the time it takes for the dissolved oxygen to equal zero. This latter time is a function of the respiratory demands of the sludge and the available dissolved oxygen. A formula to control return sludge based on this logic can be derived as shown below [1].

$$\frac{\text{time it takes to reduce DO to zero (from respiration rate)}}{} = \text{retention time of solids in final tank}$$

$$\frac{\text{DO in aeration tank}}{RR_{ml_0} - O_2 \text{ transfer}} = \frac{\text{volume of solids in final tank}}{Q_{RAS}}$$

$$\frac{DO_0}{RR_{ml_0} - 2.0} = \frac{VOL_S}{Q_{RAS}}$$

$$Q_{RAS} = \frac{1}{DO_0} VOL_S (RR_{ml_0} - O_2 tr)$$

For most conditions $O_2 tr = 2.0$ mg/l/hr. Then $Q_{RAS} = (1/DO_0) VOL_S (RR_{ml_0} - 2.0)$

Another way to eliminate anoxic and anaerobic conditions in the return sludge is to monitor the oxidation redution potential (ORP) of the return sludge. Although each plant must establish its own range of ORP values, values or ORP between +70 and -70 may indicae a possibility of anoxia and values of ORP less than -70 will probably indicate an anaerobic state.

Rate of Wasting

The normal objective of wasting is to stabilize the concentration of mixed liquor suspended solids (MLSS) in the aeration tank. A simple control strategy is then to set a concentration of MLSS and adjust wasting so that this goal is accomplished.

Various techniques have been developed to justify the "best" concentration of solids for a particular plant. It is easy to recognize that most control strategies have MLSS or MLVSS as part of the formula and therefore it is possible to solve for this value if a "best" value of the particular control parameter has previously been established. "Best" values of SVI, SRT and F/M should be established by looking at trend charts that compare a performance parameter such as final BOD_5 or SS with the values of the control parameter.

If "best" values have been established, it is easy to determine the operating level of MLSS. Following are rates of wasting based on concentration of MLSS:

Using SVI:

$$MLSS = \frac{\text{settled volume} \times 1,000}{SVI}$$

Using SRT

$$MLSS = SRT \times \frac{\text{sludge removed}}{\text{day}}$$

Using F/M

$$MLSS = \frac{BOD_5}{V(F/M)}$$

where $V = \frac{MLVSS}{MLSS}$ (estimated)

Unfortunately none of the above control parameters have any relationship to biological characteristics of the system. SVI is clearly a physical measurement of settling characteristics. Sludge age or sludge retention time is a physical measurement similar to hydraulic detention time which reflects changes in dominant species of microorganisms that might be expected. It *does not* provide any information on the viability or bioactivity of the microorganisms.

The conventional F/M ratio, in which food is BOD_5 and M is MLVSS, also does not tell much about the biological conditions of the microorganisms *in the aeration tank*. Certainly the biological action taking place in the highly diluted environment of a BOD bottle bears little relationship to the bioactivity of mixed liquor. In like manner, MLVSS is not a good indicator of *viable* microorganisms.

If the goal of wasting is to maintain given concentration of viable microorganisms then it is best to waste on the basis of a viability parameter which indicates the ratio of viability to solids. Such a viability parameter results from dividing the endogenous rate of respiration of the sludge by the concentration of mixed liquor volatile suspended solids:

$$\text{viability indicator} = \frac{RR \text{ endogenous sludge}}{MLVSS}$$

Other Controls

The use of other controls such as raw wastewater storage or microorganism storage depend on whether or not storage facilities are made a part of the total process. The effectiveness of their use is further related to the controllability of each storage facility. Clever design, using innovative concepts, is necessary to make maximum and cost-effective use of these modes of control.

LIMITATIONS OF CONTROL STRATEGIES

Control strategies cannot be effective unless three factors are present within the total plant operation. These are:

1. controllability of plant components
2. capable on-line sensors
3. a means of data management

Certainly it is futile to discuss control strategies if control of aeration, return sludge, and wasting is not possible or is severely limited.

Equally important are the sensors that supply the information used in process control strategies. The strategies discussed above utilize continuous timely measurements of suspended solids, dissolved oxygen, respiration rates, etc., and therefore sensors to measure these parameters must be reliable and precise.

Finally, it is necessary to have a well designed plan of organizing, analyzing and utilizing the data to make them most effective in process control. This could consist of simple trend plots of control parameters and effluent characteristics or an elaborate computer system which statistically analyzes all data and relates the strategy to plant performance. The goal is to utilize the "best" control strategy available under the conditions that exist *right now*. As conditions change from hour to hour and day to day, a different control strategy may become "best." Without data management the "best" control strategy may never be found or utilized.

For further information on this topic, see the Questions and Answers section on page 213.

REFERENCES

(1) Arthur, R.M. *New Concepts and Practices in Activated Sludge Process Control* (Ann Arbor, MI: Ann Arbor Science Publishers and Arthur Technology, 1982).
(2) Gray, A.C., and H.D. Roberts. "National Operation and Maintenance Survey Results Regarding Activated Sludge Process Control," in *Application of On-Line Analytical Instrumentation to Process Control*, R.M. Arthur, Ed. (Ann Arbor, MI: Ann Arbor Science and Arthur Technology, 1982).
(3) Camp, T. "Sedimentation and the Design of Settling Tanks," *Trans. Am. Soc. Civ. Eng.* (1946) p. 895.

(4) Huang, J.Y.C., "Microorganism Viability," in *Application of On-Line Analytical Instrumentation to Process Control*, R.M. Arthur, Ed. (Ann Arbor, MI: Ann Arbor Science Publishers and Arthur Technology, 1982).
(5) Arthur, R.M., L. Kent and J. Masters. "Microorganism Storage—An Innovative Answer to Excessive Inflow and F/M Control" presented at Annual Meeting Central States Water Pollution Control Association, Duluth, MN, May, 1981.
(6) Arthur, R.M. "Design of Aeration Systems—A Rational Method to Update Ten State Standards" presented at Central States Water Pollution Control Association, St. Charles, IL, May, 1979.

CHAPTER 3

SIMPLIFIED CONTROL STRATEGIES— LETTING THE OPERATOR OPERATE

Paul Klopping
    Environmental Training Consultants, Inc.
    Albany, Oregon

INTRODUCTION

Sucessful operation of activated sludge facilities requires an integrated approach involving analysis and execution of the key pressures which are available in directing and controlling this biological and physical process. The goal of operational control is the production of a sludge quality which produces the best effluent possible within the financial constraints, physical limitations and effluent quality requirements for each facility.

The dilemma facing many operators of small to medium activated sludge plants is the lack of support systems which lead to long-term, reliable process control. Typical shortcomings may include any or all of the following:

1. **Instrumentation:** It is lacking, inadequate, unreliable, or in disrepair.
2. **Laboratory:** It has inadequate work area, equipment and training for process control testing by operators; the laboratory may be reserved for effluent and compliance monitoring by "lab personnel."
3. **Data management:** It is piecemeal, and doesn't become part of "the big picture."
4. **Personnel:** The hierarchy for process control decisions is either ill-defined or lacks coordination; there is inconsistent interpretation of data and implementation of control pressures.
5. **Design inflexibility:** There is difficulty making operational changes in response to control test demands, i.e., mode, hydraulic and organic loading,

tank mixing, detention times, waste sludge and return sludge requirements, and dissolved oxygen control.
6. **Training:** There is lack of understanding and application of control tests, what they reveal, and how to implement correct control pressures.

A process management plan that deals with the lack of sophisticated support systems is one which places the plant operator in the key position of measuring and controlling sludge quality. This is accomplished through a series of simple control tests which define process status and, when interpreted correctly, reveal necessary process adjustments. The operator is intimately involved in implementing an integrated control strategy in an attempt to achieve process controllability, acceptable effluent quality, and an economically acceptable and stable operation.

Every plant must have one person (lead operator or process control supervisor) who is ultimately responsible for establishing operating parameters (targets) based on an understanding of plant design and capability and an analysis of control test trends. Within established limits, however, the shift operator can assume the responsibility for making observations, performing control tests and making process adjustments. By performing a battery of simple control tests, the operator readily acquires data which are necessary if one is to know what changes are necessary for achieving the operating targets.

The following control tests serve as the basis for a strategy which integrates the key factors necessary in defining and controlling sludge quality:

1. settleometer
2. centrifuge
3. clarifier blanket depth
4. respiration rate
5. microscope
6. turbidity
7. dissolved oxygen
8. suspended and volatile suspended solids

This battery of tests can typically be completed by an operator in three hours. Data handling will require additional time but may be greatly enhanced by the use of a microcomputer.

DISCUSSION

"Sludge quality" is a term with a broad meaning. It is as much a philosophical approach as it is empirical. At the heart of the matter, "sludge quality control" requires observations and tests which go beyond the typical control sequence of lab test, calculation and adjustment. A fourth step is included which requires the consideration of several or many tests and observations and the development of an intuitive feeling which influences the direction or degree of adjustment indicated by the basic lab test–calculation–adjustment sequence of process control.

An example may clarify how the sludge quality approach amplifies traditional control methodologies. One means of control may be a target mean cell residence time (MCRT). The operator runs a lab test to measure inventory, determines wasting rate, makes a calculation of current MCRT, and then adjusts wasting to achieve the target MCRT. The sequence of control was: lab test–calculation–adjustment.

In the same situation, an integrated approach to sludge quality control would include other factors prior to making a process adjustment. The operator would measure inventory and wasting rates, just as before. Several other observations are made to establish the status of the facility. Herein we use a battery of simple control tests to expand our understanding of the current sludge quality so that the wasting decision has a quantitative and a qualitative component.

Prior to making the wasting decision, a settleometer is run to reveal settling characteristics and the effluent quality which the sludge is capable of producing, barring the influence of the clarifier. Critical observations of floc formation during initial settling are made. Settling rate is noted, giving a feeling for the oxidative pressure which the sludge has been under. Rise time is noted so that clarifier sludge detention time can be controlled, particularly in dealing with a nitrified sludge which has a potential of denitrifying. Compaction characteristics are noted so that thickening can be translated into a return sludge flow adjustment and an optimal clarifier sludge detention time. The settleometer, therefore, is used to visualize the sludge quality. It reveals the following characteristics:

- color
- odor
- settling rate
- hindered settling/high inventory/effects of dilution
- rise time
- compaction characteristics/return sludge flow (RSF) adjustments
- effluent quality (turbidity) vs settling characteristics
- nonflocculation—dispersed growth
- deflocculation—toxicity

Sludge is examined microscopically to reveal the following criteria:

- viability
- floc size, shape, color
- filaments—amount, type
- protozoa—predominant forms

The microscopic picture is used to confirm the observations and conclusions drawn from the settleometer. It provides many clues to sludge age, floc structure, toxicity, and particularly filamentous influence on settling problems.

Respiration rates and dissolved oxygen measurements are made to evaluate the metabolic activity and integrity of the biomass. Various applications of these measurements are useful in determining how to vary the organic and hydraulic loading in an aeration tank, when the operating mode should be changed, and what sludge inventory is best in achieving the plant's performance objectives.

Solids inventory measurements are made to account for solids in the aerator and clarifier. Clarifier sludge blanket measurements are made for this purpose. The centrifuge is used as a quick measurement of solids concentration in the aerator, return and waste sludge lines. Traditional parameters can then be tracked, such as MCRT and food to microorganism ratio (F/M). Additionally, inventory measurement is coupled with the settleometer data to allow the development of a solids balance around the clarifier and an awareness of sludge retention times in both the aerator and clarifier.

Plant influent is measured for organic strength, total suspended solids concentration and mixed liquor suspended solids (MLSS) biocompatibility. These data are integrated with respiration rate observations, settleometer data and microscopic observations.

With the battery of control tests completed, the operator must interpret the data and, when necessary to achieve the established control targets, make adjustments to the controllable areas of the plant. These areas typically include return sludge flow, waste sludge flow and aeration basin dissolved oxygen (DO).

Rather than making a wasting adjustment simply on the basis of MCRT, the sludge quality operator would have looked at settling characteristics to get an idea of the oxidative pressure the sludge had been under. The microscopic observations would be considered along with respiratory activity to further get a feel for how "old or "young" the biomass was, and how it was handling the organic load applied to the plant. Flocculation characteristics and compaction properties would further direct a decision of how to control wasting to maintain current sludge quality or change it to a new biological equilibrium point. Holding a particular sludge quality may be as difficult as changing it. This is due to the fluctuating conditions experienced in influent characteristics from day to day and month to month. Integrating the above tests and observations provides a more precise means of determining current plant status—the essential step in either maintaining or changing a given sludge quality. The operator, then, makes a decision to waste an amount of sludge on the basis of inventory, its quality, and the biological requirements which must be satisfied to optimize effluent quality. Wasting decisions become much more than a mechanism of controlling inventory; they are now controlling inventory to direct sludge quality.

Adjustments of the other operating parameters, return sludge flow and air, are made in a similarly interactive way. Consideration is given to settleometer compaction characteristics and solids distribution between the aerator and clarifier. Blanket levels are considered, as are aerator and clarifier sludge retention times. Influent flow and organic strength, as well as aerator mode, are also evaluated. Respiratory activity of fed and unfed sludge is appraised, along with the microscopic observations. Return sludge flow is then adjusted to satisfy a particular sludge quality's requirements. With a solids balance established around the clarifier, return sludge flow control becomes an essential part of the sludge quality approach in establishing proper biomass distribution and oxidative pressure. Mixed liquor settleability, as measured in a settleometer, describes that particular sludge quality's compaction characteristics. If settleometer measurements are made at regular time intervals until the sludge reaches ultimate compaction (usually one hour), the concentration of the thickened sludge can easily be calculated. Return sludge flow is then adjusted to achieve a target concentration (typically that concentration where the sludge is approaching ultimate compaction). This strategy gives the operator the insight necessary to maximize sludge compaction in the clarifier while minimizing sludge retention time. Such an accommodation of the biological needs of the organisms in the system is an example of a control testing and adjustment sequence which is performed regularly by the operator, and is intended to enhance sludge quality.

On each pass through the aerator, biomass must feed on whatever happens to be in the tank at the time. Air requirements and the concentration of BOD vary, of course, and the control test regimen previously described quickly reveals bioactivity, particularly the respiration rate test and microscopic analysis. Microscopic examination of activated sludge floc usually reveals two competing groups of organisms—filaments which serve as structural backbone, and zoogleal microorganisms which develop gelatinous colonies around the filaments. Sezgin et al. [1] have reported that a dissolved oxygen gradient exists from the interior of a floc particle to the margin. To satisfy the DO needs of organisms near the center of the floc, DO targets may have to be raised as the operating F/M is increased. The textbook DO target of 1–2 mg/l may be suitable for sludges experiencing one loading, but may be inadequate to ensure DO penetration into the flocs of a higher F/M. Sludge quality control therefore requires the operator to consider organic loading in establishing targets for aeration basin DO.

The control, therefore, of wasting, return and air is most accurately achieved by defining plant status from a variety of vantage points, then using a composite approach in applying these pressures to influence sludge quality. The sludge quality approach to process control uses simple tests but represents a complex and relatively high level of responsibility for the operator. It requires some deductive thought and operating experience to put it into practice. It does, however, offer the operator the tools and understanding for making many process decisions. Particularly in small to medium plants, operators can make up for many shortcomings in design and instrumentation. With an investment in training and provision of simple testing equipment, operators can make the difference between plant success or failure.

REFERENCES

1. Sezgin, M., D. Jenkins and D. Parker. "A Unified Theory of Filamentous Activated Sludge Bulking", *J. Water Poll. Control Fed.* 50:362–381 (1978).

CHAPTER 4

UNDERSTANDING THE APPLICATION OF ANALYTICAL DATA TO PROCESS CONTROL

Thomas J. Kutcher and Gary E. Ettel
Operating Consultant Services
Cincinnati, Ohio

INTRODUCTION

A fundamental step in applying process control techniques to wastewater treatment is the understanding of the analytical data used and the assumptions made in calculating a process control parameter.

We will attempt to point out some key questions that should be asked when making process control decisions. These questions center around understanding the fundamentals of the processes and the scientific principles involved, along with knowing the limitations of the analytical data and the interfaces between man, machine, and nature.

We will first look at what a process does, that is, fundamentally. We will then discuss the selection of analytical data and what it can and cannot tell us. We will touch on process control parameters and their equations and assumptions. Finally, we will discuss implementing the changes warranted by the process control decision-making and how to follow up and monitor these changes.

We will use the activated sludge process as an example throughout this presentation to help explain the concepts presented.

PROCESS PURPOSE

Before we can think about controlling a process, we must first have a fundamental understanding of what it really does. We should be aware of the fundamental scientific principles that describe what takes place in the process. Ask yourself what really happens to the wastewater as it passes through the unit process. One way to help answer this question is to ask what is different in the wastewater leaving the porcess from the wastewater entering it. Once you can define the differences you need to ask what caused these changes to take place in the process. Finally, what factors affect that process of change?

For an example, let us take a look at the activated sludge process. What really happens to the wastewater as it passes through the activated sludge process? What changes occur in the wastewater? We add microorganisms to the wastewater in the form of return sludge, and then settle them back out again in the clarifier. What is different about the wastewater leaving the process?

1. Its 5-day bilogical oxygen demand (BOD_5) is less
2. Its total suspended solids (TSS) is less.
3. Its ammonia-nitrogen (NH_3-N) is less.
4. Its alkalinity is less
5. Its dissolved oxygen (DO) is greater.
6. It is clear, not turbid.
7. It has lost its odor.

What caused these changes to take place? It must have been the microorganisms we added. But more importantly, it was environmental conditions that we maintained that had the most effect on the process of change:

- DO level
- pH
- mixing
- no toxins
- detention time

We can say then that the purpose of the activated sludge process is primarily to reduce the amount of the soluble and finely divided organic fraction of the wastewater i.e., those materials that if discharged to a watercourse would cause an oxygen demand on the aquatic environment.

The most important item the operator needs to know about the activated sludge process is how to control the environmental conditions, and at what level they should be maintained. This though simply stated,

APPLICATION OF ANALYTICAL DATA 55

involves knowing the fundamentals of chemistry, microbiology, hydraulics, oxygen transfer and some good old common sense.

Now that we have defined the purpose of the process so that we feel comfortable about knowing the fundamentals of what is going on, we next need to look at which parameters would tell us something useful about the conditions of the process.

DATA SELECTION

When we are selecting data to be used to help monitor the condition of a process, we should first ask what we expect the data to tell us. Are we looking to find out something about the changes taking place with the wastewater's characteristics or are we looking for information on the environmental conditions of the process?

Once we identify what type of information is desired, we need to list the tests that we are aware of that can give us the desired information. We must determine how difficult it is to obtain the data; the best location to get a representative sample for what we are looking for (this may not be the same location for all tests on the same process); and lastly, whether the data will be *timely* information. How long do we have to wait to get the lab results? Is it in time to use the data for the intended purpose?

Referring back to the activated sludge process, we should ask what it is that is important to know about the process? For example:

1. How much organic matter does the system have to remove?
2. Did the system do the job?
3. What are the environmental conditions?
4. What changes, if any, should we implement?

Knowing that this type of information is desired, let us see if we can list the tests that we are aware of that could give us the desired information:

Types & Quantities of Organic Matter

- BOD_5
- chemical oxygen demand (COD)
- total organic carbon (TOC)*
- volatile suspended solids (VSS)
- flow*
- oxygen uptake rate (OUR)*
- total Kjeldahl nitrogen (TKN)
- NH_3-N*
- toxins*

Monitoring these parameters entering and leaving the system will tell us to what degree we have accomplished our goal. Note that the items marked with asterisks can be monitored continuously with on-line instrumentation.

The following tests can tell us something about the environmental condition, from which we can predict how the process is performing:

Environmental Conditions

- temperature*
- pH*
- DO*
- mixed liquor suspended solids (MLSS)*
- mixed liquor volatile suspended solids (MLVSS)
- solids settling rate*
- alkalinity
- nutrients
- OUR*
- viability (living microorganisms)*

Since the activated sludge process is a living system, we need to concentrate on data that will tell us something about the changes taking place, and not just static measurements—i.e., numbers at one instant in time. We need to know not only what is changing, but at what rate and in what direction.

Other important questions to ask are: What are the limitations of the testing procedure? What are the similarities between the testing conditions and the actual process conditions? What about interferences? . . . sample degradation? What is included in the end result? What is not measured by the technique but still reacts in the process and must be considered? Be aware of the type of information that the type of sample you collect will give you. Grab samples can give you data about what is going on at the instant of sampling. Composite samples, on the other hand, can give you average data over the period of compositing, provided degradation has not taken place. On-line analyzers provide still another type of information—namely, instantaneous data with continuous output. If such on-line analyzers are connected to a microprocessor or computer, trend analysis can also be obtained. In other words, you get the best of both worlds.

Before collecting the data we should write out how we are going to use the data. We could intend to use it in the following ways:

1. process control
2. enviornmental conditions monitoring

3. historical information
4. coincidental information

This leads us to the next topic—data utilization.

DATA UTILIZATION

What will we do with the data once we obtain them? We will probably first transfer them from a lab worksheet to a summary form for the plant or process (usually these are set up to contain a month's worth of data). We will plot the data on some sort of trend chart to give us a better perspective of what it means, especially as it relates to past data. Lastly, we may use the data to calculate a process control parameter, and may then again plot the control parameter on a trend chart.

Figure 1 shows a typical worksheet for calculating BOD results. The average values for the raw influent, primary effluent and final effluent can be used to determine how the activated sludge process is doing. First, these values are transferred to a monthly summary form (Figure 2). These values are used to calculate percentage removals, organic loadings, F/M ratios and the like. These values can then be trend charted on specific graphs selected by the operator to give a view of how things are changing and how changes that were made affected the process [1] (Figure 3).

PROCESS CONTROL PARAMETERS

How do you pick a process control parameter for your plant and process? First, list all the process control parameters of which you are aware. Check reference books and journal articles you may have. (For example, Stall and Sherrard [2] wrote an article on their investigations evaluating control parameters for the activated sludge process.) Also ask around to see what other operators are doing and what success they are having. As can be seen from the following list of activated sludge process control parameters, there are many different methods of controlling the activated sludge process.

Activated Sludge Process Control Parameters
- food to microorganism ratio (F/M)
- sludge age (using TSS, BOD, COD)
- Mean cell residence time (MCRT)
- constant MLVSS

58 ACTIVATED SLUDGE PROCESS CONTROL

SAMPLE: TESTED:	BLANK	RAW INFLUENT		PRIMARY EFFLUENT		FINAL EFFLUENT	
1. BOTTLE NO.	46	47	48	50	51	52	56
2. % DILUTION	100	98	97	96	94	80	70
3. % SAMPLE	0	2	3	4	6	20	30
4. SAMPLE DO	8.9						
5. INITIAL DO_i		8.7	8.6	8.5	8.3	8.0	7.8
6. 5-DAY DO	8.8	5.2	2.8	4.5	1.6	4.8	1.8
7. DIFFERENCE	0.1	3.5	5.8	4.0	6.7	3.2	6.0
8. $(B_i - B_5) \times \%$		---	---	---	---	---	---
9. DIFFERENCE		3.5	5.8	4.0	6.7	3.2	6.0
10. BOD_5, mg/L		177	193	100	112	16	20
11. AVG. BOD_5		185 mg/L		106 mg/L		18 mg/L	

NOTES:

Line
2   % dilution equals amount of dilution water to the BOD bottle, expressed as a %.
3   % sample equals amount of sample added to the BOD bottle, expressed as a %.
4   Sample DO is the mg/l of dissolved oxygen in the blank and samples before setting up the BOD.
5   Initial DO_i is the measured DO of the dilution water/sample mixtures. (It is measured 10 to 15 min after the samples are set up.)
6   5-Day DO is the DO after 5 days of incubation.
7   Difference equals Line 5–Line 6.
8   $(B_i - B_5) \times \%$ equals the amount of DO used up in the dilution water in the BOD bottle containing the mixture of dilution water and sample, and equals Line 7 of "Blank" column × Line 2 of "Sample" column.
9   Difference equals Line 7–Line 8.
10 BOD_5 equals Line 9 divided by % Sample in Line 3.
11 AVG. BOD equals the sum of the two BOD values obtained for a sample divided by 2. (Note: If one of the BOD values is way out of line, disregard it and use the other value.)

Figure 1.   BOD worksheet.

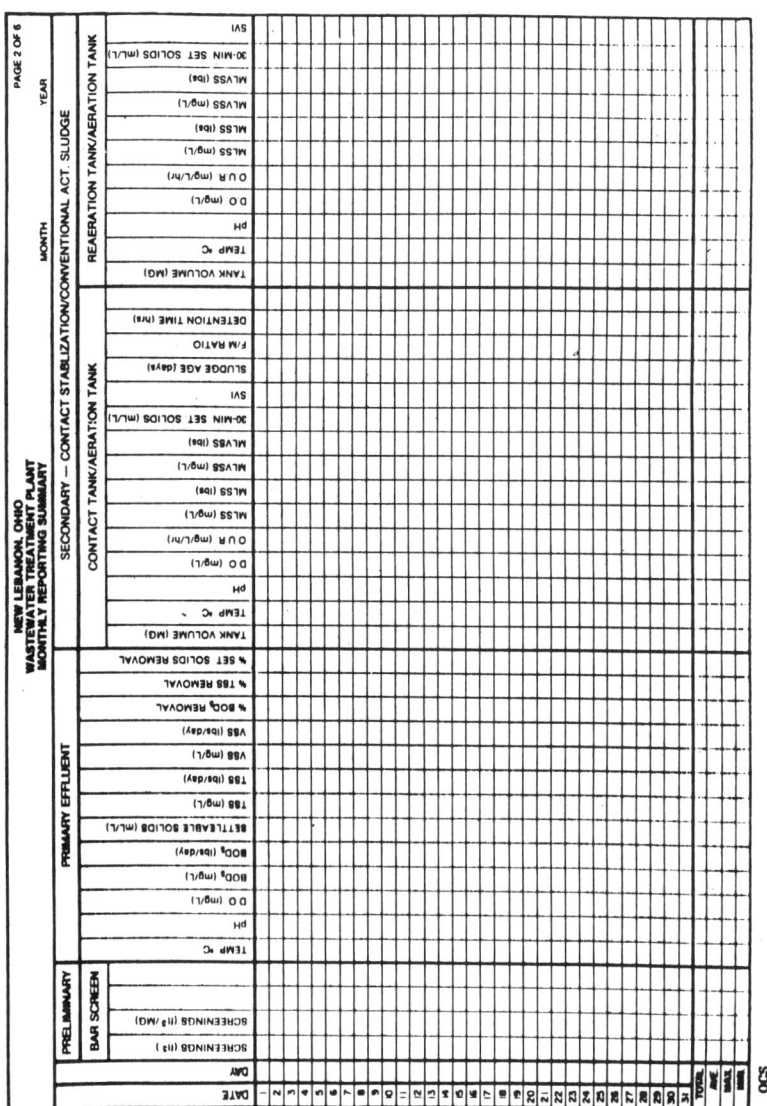

Figure 2. Monthly summary form.

60 ACTIVATED SLUDGE PROCESS CONTROL

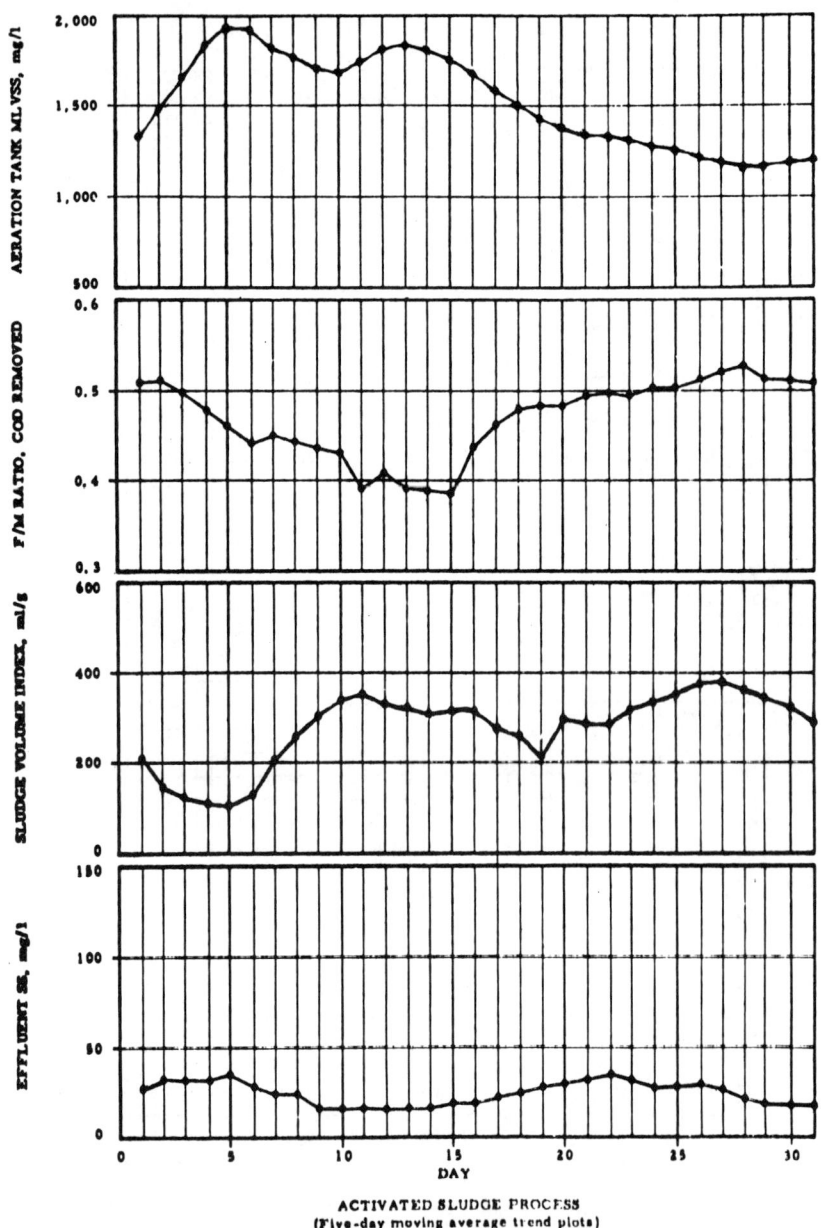

Figure 3. Trend charts.

- biological food to volatile microorganism ratio (F_b/M_v) (using respiration rates [3].
- microscopic exam
- DO level control
- settleable solids control
- return sludge control
- sludge quality control (Al West method [4])

In most plants, a combination of several of these methods would be utilized because each one tells the operator something slightly different about what is going on.

In evaluating your selection of process control parameters, ask yourself what a particular process control parameter is telling you. This can be done by looking at the component parts used in the calculation and analyzing each component separately. You may wish to rate each component, on a percentage basis, as to its ability of giving good, accurate information on which to base decisions. What is the accuracy expected? What are the limitations expected? Then, to find out how good the process control parameter will be for your application, simply multiply the decimal percentage ratings of the component parts (e.g., F/M: if F is rated at 85%, and M at 75%, then the F/M rating would be 0.85 x 0.75 = 0.64 or 64%). You should ask yourself if there are ways to make the component parts more acurate for your situation. Are there new test procedures, new instruments, better sample locations, more frequent sampling or analyses that would improve the quality of the data being used, thereby improving the effectiveness of the process control parameter being utilized?

It is also important to know what assumptions are being made for a particular process control parameter. Try writing them out as you see them and then ask yourself whether they apply to your situation. If not, can the equation be modified or weighted to give it more meaning for your situation? If it cannot, then do not use it but simply keep track of it awhile and compare it to other process control parameters you are using to see if it is sensitive to your process changes.

Let's look at the assumptions made for the classical F/M ratio—i.e., the pounds of food (BOD_5) divided by the pounds of microorganisms (MLVSS). The first assumption made is that a pound of microorganisms can consume "X" pounds of food. Secondly, it is asumed that the BOD_5 test gives an accurate representation of the quality and amount of food entering the system. Likewise, the MLVSS test is used to approximate the number of microorganisms. Researchers [5, 6] have shown that the amount of living microorganisms compared to the MLVSS can range from 10 to 60%. The environmental conditions are so different in a BOD bottle from those in the aeration tank and the fact that the test takes five days to obtain results is one reason that this classical method of control, while well intentioned, has not been a very practical control parameter for operators.

Arthur [7] has introduced a new food to microorganism ratio concept. Biological food (F_b) is measured by the respiratory activity of a mixture of return sludge and the incoming food, and viable microorganisms (M_v) are measured by the rate of endogenous respiration of the microorganisms. This is definitely an interesting concept because both F and M by this definition are capable of being measured continuously with on-line instruments. Only time will tell how sucessful this control scheme will be, but for the first time an operator will have data available that can be utilized to anticipate what will happen to the process. We will not have to wait to see what goes wrong and then react to the best level of our abilities based on the art as we understand it.

We certainly hope that with inventions of better process monitors, on-line analyzers, microprocessors and computers that we in the operating profession are on the verge of a new frontier in wastewater treatment. It may take a generation or so to come to full realization, but someday things like sludge age and MCRT may be antiquated as Imhoff tanks and manual cleaning of bar screens and grit tanks. The goal of a process control program must always be to use the most accurate process control parameters that give timely information with the least amount of effort.

IMPLEMENTING CHANGES

No matter how accurate the process control parameters may be in telling us what is going on, you and your equipment must be able to respond to the changes that are required. This means that you should know what you can and cannot do with your system. Consider what the effects on other process parameters might be before making a change. You could make a change based on one process control parameter's results and end up with a worse situation than when you started. We must also respect Mother Nature—she will not let us do everything we want to do. She has laws, you know—and of course we cannot forget Murphy's laws either.

Let's look at the chain of events that takes place in a process control program:

1. A determination is made through a process control parameter that something is out of line or needs adjusting.

2. The required change has to be communicated to others. Beware of the loss in the exchange of communications. Try having the communicatees repeat back to you what you told them and then ask them why they are making the change. If they do not understand why, then take the time to explain it to them. It is good insurance for getting the job done right the first time.
3. The process equipment is changed in some way. Make sure you know whether your equipment can do what you re asking it to do. None of us has a custom designed treatment facility. We group pieces of standard equipment together to make up our treatment facility and most of the time we must accept what is given to us. Thus it is important to know the limitations of the equipment.
4. We wait for nature to respond to the changes we have made. We can only practice indirect control and patience. We control the process environment through the process equipment, which should in turn allow nature to respond accordingly to its laws as we understand them.

This approach is called *reactive control*, in which the operator reacts to things as they occur and are discovered. Utilizing the newer on-line analyzers, microprocessors and computers we can attain a level of *anticipatory control*, in which we monitor the rate of change of the process performance and rate before a problem can occur. It is felt that this will be the method of control for the future since it is capable of maintaining a much higher effluent compliance percentge.

PROCESS MONITORING

Wastewater treatment is an art and not a precise science, and because of this we must continue to monitor the process from day to day. Not having control over or not even knowing what comes into the plant also makes the job challenging. But there are signs of hope out there. The more data we collect and analyze the more we learn the beauty of the world in which we live and operate. Every now and then, one of us is blessed with a gift of insight that opens up a door to another whole world of knowledge. Use your time wisely. Think before you act. Ask questions of yourself and others about what you do and see—it is a sign that you are a caring operator—it is not a sign of ignorance. Wastewater treatment technology and the operating profession have advanced because of this very simple principle.

REFERENCES

1. "Process Control Manual For Aerobic Biological Wastewater Treatment Facilities," U.S. Environmental Protection Agency, Manual No. EPA-430/9-77-006 (Washington, D.C.: U.S. Government Printing Office, 1977).

2. Stall, T.R., J.H. Sherrard. "Evaluation of Control Parameters for the Activated Sludge Process," *J. Water Poll. Control Fed.* 50(3):450-457 (1978).
3. Arthur, R.M. "The Use of On-Line Respirometry in Control of the Activated Sludge Process," AIChE 71st Annual Meeting, Miami Beach, Florida, November 12-16, 1978.
4. West, A.W. "Activated Sludge Quality and Process Balance (DSA & BLT versus CSP)," WPCF 51st Annual Meeting, Anaheim, California, October 1-6, 1978.
5. Blok, J. "Measurements of the Viable Biomass Concentration in Activated Sludge by Respirometric Techniques," *Water Res. J.*, 10:919-925 (1975).
6. Walker, I., and M. Davies. "The Relationship Between Viability and Respiration Rate in the Activated Sludge Process," *Water Res. J.*, 11:575-578 (1976).
7. Arthur, R.M. "On-Line Measurements Improve Activated Sludge Process Control," *InTech*, (September, 1980), pp. 103-106.

CHAPTER 5

DATA MANAGEMENT FOR PROCESS CONTROL

Owen K. Boe
    Envirotech Operating Services
    San Mateo, California

INTRODUCTION

Wastewater treatment management often has received less attention than it deserves. Often we as consultants find ourselves responding to only upset situations. The solutions and recommendations left behind, therefore, may only address abnormal conditions. Full-time management needs to address status quo conditions and must provide mechanisms and systems to prevent future upsets.

A major concern of any manager is how are people going to respond and how timely they will respond to a given situation. The general management concepts are universal to any type of organization and are summarized as follows:

Management Concepts
- organization
- communication
- delegation
- control
- feedback
- forecast

The application of these concepts, however, may be difficult to describe and define in every organization but their application is essential for efficient and effective plant management.

DATA MANAGEMENT APPROACH

Data management can be referred to as a building block for plant operations. This is absolutely true, because without proper, accurate and precise data, plant management becomes a guessing game. We probably have all experienced situations where we have been party to or have seen decisions made based on erroneous data. On the other hand, data management and data collection can become an exercise that drives many plant operators, lab personnel and others totally insane. A key aspect of data management that is sometimes, and unfortunately, overlooked is providing a close examination of how the data are to be used, when they are to be used, where they are to be used, and by whom. Without these considerations, the collection of data often becomes an exercise just for the sake of collection.

I refer to the situation just described as the result of a bottom-up approach. The bottom-up approach can lead to many unanswered questions and management inefficiencies, so let me define this approach one step further so that its shortcomings can be clearly understood. The bottom-up approach, in its worst case, is a program that is designed in an office for office use and by office personnel. This approach looks at a treatment plant and a series of circles and squares and defines all the various parameters in their locations that are needed to understand entirely what goes in and out of those squares and circles.

The bottom-up approach has great merit for doing carefully controlled research studies, engineering analyses, and for determining quantifiably the relationship between the circles and the squares. Unfortunately, this is not the typical or the normal situation at most treatment plants. Figure 1 depicts a typical bottoms-up approach by representing the flow of data toward management. The burden of data interpretation rests entirely on the manager; this results in ineffective use of the organization. The plant manager in this mode is overloaded with data and is required to synthesize data himself to gain meaningful information. This takes time and, most important, it takes him away from other management activities.

Since the objective of treatment operations is to produce an acceptable quality effluent in the most cost-effective manner, the plant manager's time is essential for decision-making, not data analysis. If analysis and synthesis of data can be made by other levels of plant management, then

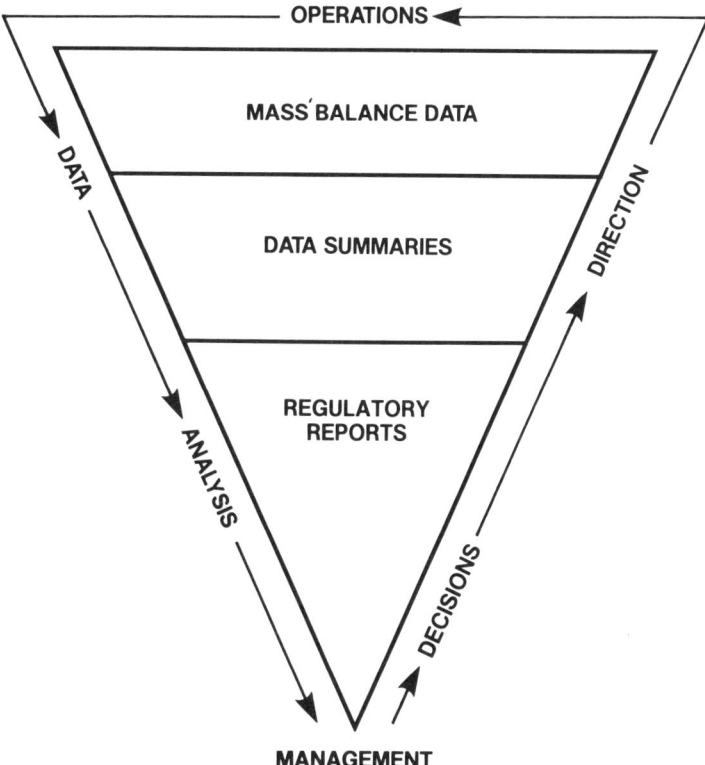

Figure 1. Bottom-up data management.

efficiencies can be realized. To be effective, however, roles, expectations and needs must be defined explicitly in order for the staff to better support management. This is now what I call the top-down approach to data management.

Figure 2 depicts this approach as a pyramid where data are reduced to information before they reach management. In this mode, the manager may never see data unless problems occur. What he sees instead are flags, targets or exception reports when a process has deviated from normal. Further, to support this approach, I define a process control plan as being an organized approach whereby the data needs and responsibilities are defined for the various operators, supervisors and management personnel.

The execution of the various levels of the plant organization are completed more smoothly when the organization of data flow has been

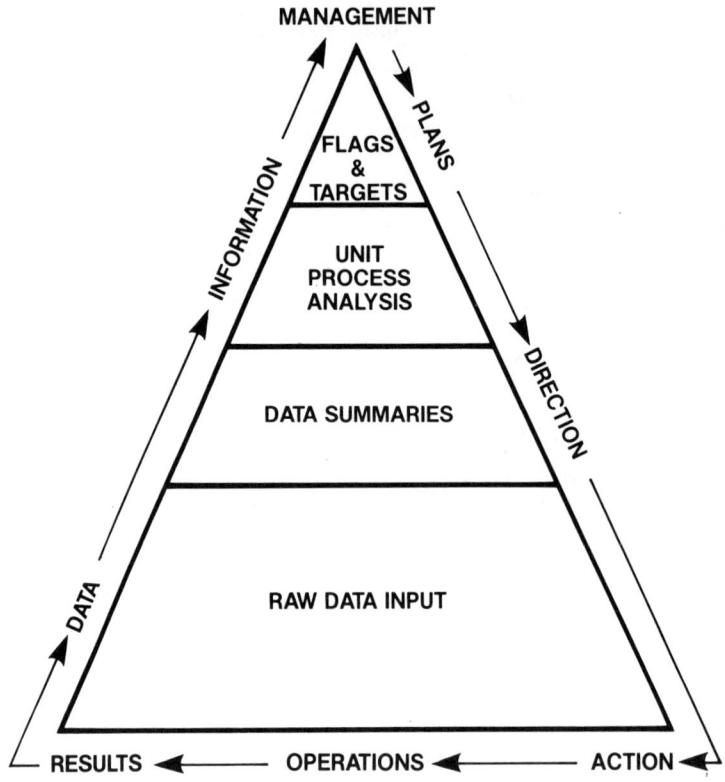

Figure 2. Top-down data management.

carefully defined. Who benefits from this organization? All layers of the plant management benefit. The operator who sometimes is low man on the totem pole benefits from having a better understanding of where his input fits into the overall program. The plant supervisors benefit because they work as the middleman between the operators and management and, consequently, they are responsible for coordinating the input and the output of the program to the plant operators. They also make decisions that govern the action of many other personnel. These decisions have to be consistent within the overall structure and direction that higher management has selected. The plant manager needs to have asssurance that the program he has defined is going to be executed smoothly, efficiently and according to a plan. If he has to devote his time and energy to reviewing all the raw data that are accumulated throughout the cycle of a daily program in a treatment plant, he is not going to be as effective in the area of planning and managing the future changes needed in the treatment plant. However, he still needs a certain select flow of information that he can review to track the major elements of the treatment plant processes, and he needs to know that additional information is available if he wants to pursue a problem.

PROCESS CONTROL PLAN

A process control plan is the plant manager's tool to achieve the ultimate goal of meeting effluent quality standards through organizing the combined efforts of a plant organization. By definition, therefore, a management plan is a top-down structure. That is, one person sits at the top (plant manager) and information has to flow up to that person from the organization. In turn, definition and direction have to flow back down the organization to the various people that direct and implement activities.

Objective

The objective of a process control plan should be to provide the mechanism, the structure and the organization of the flow of process indicators such that the manager can determine if the unit processes are in control. It is important to realize here that managing unit processes by monitoring the effluent quality alone is not a control program. At that point serious trouble already exists and prevention steps are limited; that is why "leading" process indicators are essential for management control.

Goal

The goal of the plan should be to ensure that a minimum system of uniform process control management procedures for a particular treatment plant are maintained.

Elements

The process control plan should include specific consideration of the following process management elements:

1. Assignment of process control roles and responsibilities:
   - organizational chart
   - key staff process control job descriptions
2. Maintenance of formal process management communications:
   - process control meetings

- key parameter status
- unit process reports and analysis
- detailed tabular data reports
- daily process control directives
- operator check sheets and logs

3. Unit process strategies defining a detailed management approach to each process, including:
    - process objectives
    - key control parameters
    - key parameter alarm levels
    - process control responses
    - minimum process data base
    - contingency plans
    - operator log and check sheets

Process Control Recordkeeping

All process management information (i.e., alarm reports, unit process reports, process control directives, tabular reports and graphs) should be maintained in comprehensive process control notebook(s).

Management Concepts

The process control plan should utilize management control concepts which include flags, targets, trends and forecasts, and report by exceptions. These concepts are briefly defined as the following:

Flags

Flags should be predefined for various process elements so that management can be directed to problems as they are developing and not after they occur. Flags should be defined with varying degrees of severity, such as either "warning" or "alarm."

Targets

One of the powerful tools available for process management is in the approval and evaluation of targets. Process directors and supervisors should be responsible for target recommendations, but management should provide review. The key interchange occurs when the status (actual value) is reviewed against the established target. If the target value is not achieved, then the manager should ask for reasons and for recommendation on how to achieve it. In some cases, the appropriate action would be to change the target, but this will occur only after justification has been provided.

Trends

Trends are defined as the change from one time frame to another. A linear regression analysis can be used to calculate the slope of a trend line, or the slope of plotted data can also be used to determine a trend. Review of trend information enables management to anticipate problems or improvements that may be coming.

Forecasts

Forecasts are defined as a prediction of a parameter at some time in the future. Activated sludge plants generally run with MCRT of seven days, so a weekly forecast which anticipates a process change is generally a minimum goal. Process control directors should try to forecast key variables. These forecasts can then be translated into targets for selected parameters.

Report by Exceptions

In larger organizations, a formal report by exception provides management a mechanism to learn about any specific deviations from predefined conditions. This effort removes ambiguity of when and what to report. It also reinforces the responsibilities of various individuals to manage the day-to-day routine (not exceptional) situations.

PROCESS CONTROL ORGANIZATION

In all but the smallest plants, the tasks essential to effective process control are performed not by one individual, but by an organization comprised of many people. The first (and most important) activity in a top-down approach to process management is that of assigning process control roles and responsibilities within a staff structure organized to provide well defined lines of reporting and accountability.

Plant Organization

Day-to-day responsibility for process direction should be consolidated through one person—the process control director. The process control director should also supervise the implementation of process control decisions.

Process Control Job Descriptions

Plant management assigns process control roles and responsibilities to specific individuals via job descriptions. The functions essential to process control fall into five major categories:

1. process control management
2. process control direction
3. process control analysis
4. process control implementation
5. process control data collection

Figure 3 summarizes a typical functional organizational structure. Functional assignments may overlap into various job positions. This is not a problem when it occurs, but it reinforces the need to have assignments and responsibilities clearly defined.

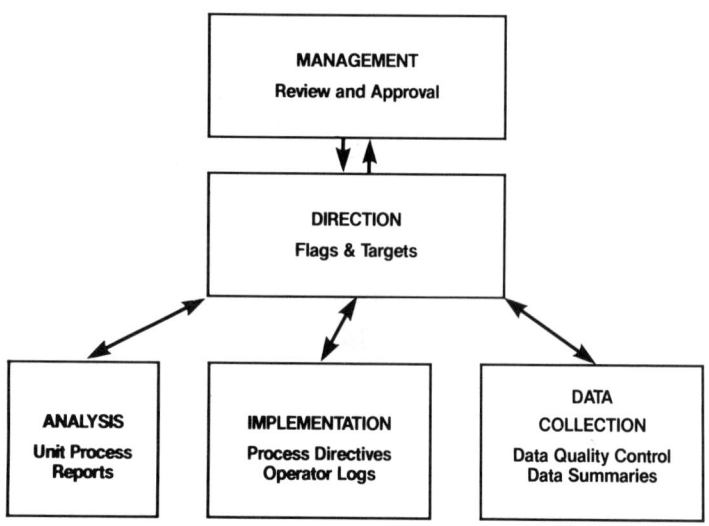

Figure 3. Process control functions.

MICROCOMPUTERS IN DATA MANAGEMENT

Envirotech Operating Services (EOS) has specialized in treatment plant management for over ten years now and currently manages fourteen treatment plants under full management contracts. Three years ago EOS started investigating the benefits of small computers to support and enhance plant management. One of the first areas explored in detail was that of process control.

Due to the contractual nature of EOS's work, economics was a key consideration. This consideration translated into the development of a computer system that was economical in both hardware and software. The result of this effort has produced a microcomputer system that is now currently installed in seven EOS treatment plants.

Each system in EOS [and typical for plants of 1–50 million gallons per day (MGD)] includes a video CRT display, a terminal and two floppy disk drives. The technical aspects of these units is beyond the scope of this chapter, but Figure 4 provides a schematic of EOS's system layout.

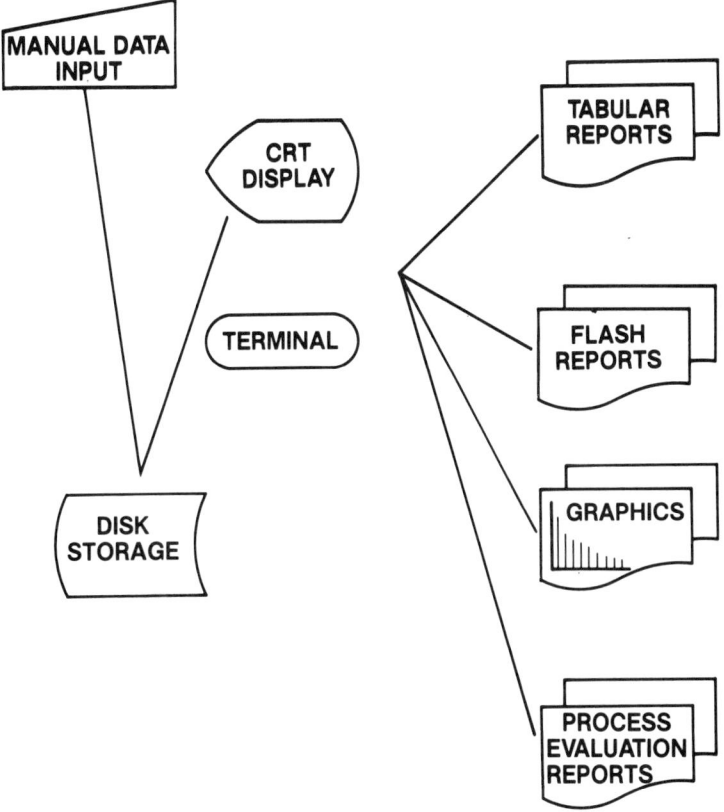

Figure 4. Cameo process control data management.

COMPUTER-ASSISTED PROCESS MANAGEMENT

The microcomputers that are offered today provide considerable flexibility and power for aiding data management. A key idea here is aiding or assisting plant management. If a computer is to be used only as a complicated and expensive calculator, then very little of the computing power is utilized and one should question whether or not a computer is more cost-effective than a good programmable calculator. On the other extreme, it must also be recognized that these computer systems are not designed for providing on-line direct control of unit processes. Instead, these systems, when properly designed and applied in treatment plants, can provide for powerful management tools which support existing management goals. In the case of EOS, the computer programs were designed to support the goals and objectives of EOS's process control plans. In this context, there are five major elements in the computer program which are essential to support the goals of the process control plans. To be consistent with the top-down approach as described earlier, these elements of the computerized package are prioritized in the following sections.

Flag and Target Status Reports

If top level management is to be protected from the burden of screening and evaluating large quantities of raw data, then the data must be reduced to meaningful flags and targets. A serious management consideration is in the leveraging capability of evaluating a few properly selected parameters to provide a quick and effective grasp of the overall program. A second advantage of this consideration is that it allows the staff, or more specifically, the process control director, to exercise his responsibility for managing the details of day-to-day or week-to-week process control. The microcomputer fits this philosophy very nicely in that it provides the memory, the calculating capability and the storage capability to evaluate selected parameters against predefined flags, which helps in the setting of process targets. Used in this mode, the computer provides a very key management report that may involve as few as 20 parameters from a plant data base that could exceed over 200 parameters.

Unit Process Reports

Process control by nature requires technical input. In this context, there is a need to look at unit processes in a fairly detailed and analytical approach. Again, a microcomputer can effectively organize selected parameters by the various unit processes and perform statistical analyses of these parameters. In this approach, various unit process reports can be generated that provide a more in-depth review of that process status, the history of that process, and even the ability to trend or forecast the direction of various parameters in that unit process.

Data Plotting

Data plotting, along with the more technical unit process reports, is another obvious advantage that microcomputers can provide to support plant management. Most operators and plant managers have, at one time or another, spent considerable time plotting and graphing various data points on paper in the attempt to explain to someone else what is happening in the process or to better understand the various relationships and events that are happening for their own analysis. In this context, data plotting on a CRT or on paper becomes a very powerful tool for aiding process analysis and process interpretation. Rapid graphical representations of data in a properly programmed computer system can also be used to provide further statistical analyses of that data.

Data Reports

Tabular or raw data reports are historically a burden that is imposed on many plant operators. These reports are sometimes used as a guide for process management, but generally the effort that is required to complete these reports usually puts their availability way beyond the time frame in which they could be useful. Secondly, the manpower required just to move data from one worksheet to another worksheet and possibly to a third or fourth worksheet can be considerable. The microcomputer, again, is an excellent tool for organizing data and various data summary reports that can be used to fulfill NPDES reporting requirements, city hall reporting requirements, or other data reporting needs.

Data Quality Control

The microcomputer used in process control also provides or can provide data quality checks. To manipulate any data, they have to be inputted into the system. Through data input software considerations, data validation checks can be designed into the system. This is especially critical when using computers, because through the means of bulk data storage it is possible to repeatedly utilize a piece of datum that is erroneous or improperly measured, monitored or entered. In this context, it is very important to have a series of validation checks for confirming the validity of the data. Again, the computer power can aid by checking the data against predefined normal ranges or expectations.

SUMMARY

Data management has often been delegated to either consulting engineers or to laboratory technicians. In either case, the interests of the key player—the plant manager—have often not been met. If data management needs are first defined as a mechanism to make a plant manager more effective, then plant operations and process control will become more efficient.

The key to providing a rational data management program is in preparing a process control plan. The process control plan should provide the structure and rationale for handling the data needs at a treatment plant.

Finally, when a concise, rational and well defined plan exists, a microcomputer provides a powerful tool for plant management. The benefits received should result in more efficient and effective use of all plant personnel for providing economical wastewater treatment.

For further information on this topic, see the Questions and Answers section on page 215.

CHAPTER 6

COMPUTERIZATION— CAN IT WORK FOR YOU?

Robert A. Davis
Thorn Creek Basin Sanitary District
Chicago Heights, Illinois

BACKGROUND

Wastewater treatment plants are very complex by nature. They must handle a raw product of unknown quantity, quality and characteristics and treat it through mechanical, physical, chemical and biological processes.

The water pollution control industry has been historically behind other process control industries. High-technology industries, such as the petrochemical industry, have been precisely controlling processes through the extensive use of intrumentation and automation for years.

Treatment plants, on the other hand, have been built as oversized processes with redundant equipment which have only minimal control. These plants would more or less be left alone and, other than mechanical failures, they would continue to operate.

This is no longer the case. The luxury of being able to put in extra equipment or tanks is long past. Due to the rising costs of construction, labor, chemicals and power, the new treatment works are being designed and sized much more precisely.

These new plants are also required to produce a much higher quality of effluent. Water quality standards, as well as ammonia and phosphorus removal, are becoming the normal goals of many plants.

To achieve these goals, and do so at a minimum of costs and a maximum of consistency, pollution control plants are adopting the exact process control techniques utilized in other industries.

The techniques, however, require much modification due to the ever changing nature of the raw sewage influents and due to the complexity of the treatment systems. A chemical industry is working with pure raw products at carefully controlled environmental conditions. This is not always possible at a wastewater plant.

The liability of the operator is somewhat different also. The chemical plant operator can cost his plant money and himself his job, but a water pollution control operator, in addition to these costs, can cost his community it's environment and find himself in legal trouble.

As a result of all this, the new publicly owned treatment works (POTW) are becoming very complex, with distribution networks, sensitive processes, flow equalization, covered tanks in pure oxygen environments and, most of all, a highly skilled technical staff.

The extensive use of intrumentation and automation is almost a mandatory requirement of these types of control techniques. One needs to know the conditions present in a process at all times to make proper process control decision.

SPECIFICATION AND INSTALLATION

Digital electronics are the tools which are used to gather, analyze and even change the vast amount of information required. These electronics can be in the form of distributed microprocessors, a centralized computer system or a combination of both.

At the Thorn Creek Basin (TCB) Sanitary District the answer in 1975 when the advanced treatment plant was designed, was the inclusion of a data-logging computer system.

The computer system required was specified to be able to collect information from the entire treatment plant and be able to provide process control functions at a future date. It was also specified to provide accounting and user charge billing functions.

The suppliers who bid stated that the accounting functions were not compatible with a process control computer and would therefore bid two separate systems. The accounting system was rejected and TCB went with the plant computer only.

It was very important when purchasing the computer to have general specifications which designate the type of functions to be provided and the future capabilities and not exact technical descriptions of equipment and their functions. This allows systems which do similar things in different ways to be competitive by meeting specified goals in their own way.

The expandability and flexibility for future requirements is a key issue. To fully utilize the capabilities of a digital processor, you must be able to grow with it and continually update it.

The installation of the system has helped TCB staff to realize some problems and benefits. In the new advanced treatment facilities the installation was much easier than retrofitting the old plant. The conduit schedules can take into consideration all of the field wiring, the location and types of meters can be assured maximum compatability with the computer and any equipment can be connected much more easily. It also showed us that it can be overdone, especially when you're designing a new plant. It is very easy to connect everything to the computer for remote control and to attempt to meter every variable condition. However, some equipment does not lend itself to remote control, and too many alarm signals or faulty meters only complicate things and make it harder for an operator to determine what is really critical.

The old secondary plant was more difficult to retrofit, but a tunnel system that exists between buildings did allow the conduits to be run to the computer with a minimum of excavation.

Another recommendation in specifying and installing the field wiring is to always pull full cables even when only using a few pairs of wires. Eventually you will probably want the extra signal capabilities and it is very expensive to keep laying new conduits. Complete point-to-point as-builts and electrical supervision should also be included in your specifications or engineering contract.

The other part of the entire installation is just as critical as the field terminating and the computer itself. That part is the software. The original software was specified to be able to scan data, compare it against alarm limits, scale it, offset it and organize it into preprogrammed formats and conveyed to output devices or memory. It was also designated to have the ability to provide several types of logging reports and display functions and be able to accept specified numbers of analog and digital inputs. The ability for the operator to modify the operating programmed portion of the software was also built in, as well as the hardware necessary for the input programming. Many installations lack the input devices needed, such as card readers and tape drives, thereby limiting the future utilization of the system severely.

The supplier then contacted TCB and requested a list of inputs, their description and an identifying name, as well as the information desired on the logging reports that were specified.

Training for three key personnel was also provided. After the initial two-week training period and receipt of copies of the operator-oriented software, TCB personnel modified the data base and rewrote much of the logging programs. The plant personnel also compiled all of the graphic displays for the plant and subsequently made further modifications to the software.

This resulted in a duplication of effort by the manufacturer's programmers and TCB staff. If a plant is sufficiently complex to warrant a computer installation, it should also have trained personnel on its staff who can program and interface with the system directly as opposed to a certified operator's routine communication with the displays and logging reports.

To fully utilize the system's capabilities, the owner must understand his custom software and be able to update, modify or expand it without having to go to the supplier each time. Software must be routinely maintained, just as a piece of equipment is. There is no way that you will not have to contact the supplier for problems with the internal software that makes the system work because it is more than likely a team effort to put an entire system together and it is also the result of many years of continuous programming and expertise.

The part of the software that is uniquely yours is another matter. The owner should know his own data base, logging reports and control logic. This also brings about another part of the design specifications which ensure the system's flexibility and future application. Not only should extra wires and terminals in the computer be provided but the space in the software must also allow for growth and additions.

In 1978, bids were let for excess flow facilities and also to expand the computer system and include some direct control by the computer. The memory of the central processing unit was expanded and field interfacing added with allowed direct digital control. As a result of these expansions, the entire software had to be rebuilt. This time TCB told the manufacturer to put the software together for a specified number of inputs and outputs and a certain number and size of logs and control programs which would all allow for a reasonable amount of growth. TCB staff then programmed the entire data base, logging reports, and an activated sludge simulator and is continuously programming and modifying supervisory control programs.

An increased number of training periods was also specified at a rate provided by the supplier so that any unused time could be credited to the contract. This training consisted of internal system software classes and could also include hardware maintenance sessions.

The main thrust of this latest expansion, besides being able to accommodate the new facilities constructed, was the process control capabilities.

PROCESS CONTROL

Process Control through the computer can consist of retaining manual control, but doing so with better information. The speed, quantity and accuracy of the information can be much improved through the data processing system.

The direct control through the computer can be accomplished in several ways. You can have operator-directed control (ODC) where the operator sits at the console and controls the data base through the various displays. The control can be direct digital control (DDC) through a properly defined data base. This is similar to what a programmable controller provides. The control can also be through custom written supervisory programs or program directed control (PDC).

The supervisory programs can manipulate the data base and make logical assumptions or choices and can communicate with the operators or programmers as much as desired. Variables can be programmed, in that can be set through displays, and the programs can communicate with each other in complicated networks.

Direct control at TCB can be performed on several different processes.

The use of computer logic for flow equalization and increased pump station efficiency is a very beneficial ability. Cost savings have been realized through energy savings by reducing pump cycling and by operating at the maximum efficiency on the pump curves.

The use of blower control through variable-speed blowers and dissolved oxygen (DO) meters has a great deal of promise. One of the intial problems encountered was that the air valves at each of 12 submerged turbine aerators has to be reset for any air flow change. A piping modification, motorized valves and/or pressure transmitters are all under consideration. Further work is required before full remote control can be effectively utilized.

The district is exploring the possibilities of helping the nitrification system accommodate large flow increases through modeling and simulating the system, varying the parameters theoretically. The control of wasting rates is also possible with some minor field modifications. Control of return sludge has only minimal possibilities due to the physical limitations of clarifier design.

A network analysis and flow calculation program has proven most useful. The TCB plant consists of three parallel primary plants followed by two parallel secondary activated sludge plants and then a tertiary nitrification activated sludge system, sandfilters and postaeration. The flows from each of these processes can be diverted, mixed and bypassed in many different nodes. By analyzing flow values and gate positions, meters can be activated or their calculated values stuffed into the appropriate software slot.

Digester performance has been significantly improved through better information. Thermocouples provide temperature data from the sludge heat exchangers. The digester gas flow to the heaters or the wasting torch are the only methods of measuring gas production and, combined with the gas pressure measurement within the system, allow maximum utilization of the gas produced. The rated capacity of the anaerobic digestive system has been increased by 20%.

Watt-hour meters measuring the utility consumption have allowed the use to be trended and wasteful practices identified. Procedures have been shifted so that the peak usage periods can be shaved.

Other control capabilities have not proved themselves as yet. Pumping the raw sludge out of the primary clarifiers has not been accomplished successfully due to several factors. The remote control of sludge valves has been very difficult due to plugging of the valves, which then trip out. Their limit switches require precise setting and fail on occasion. Another problem is that the sludge pumps can become air-bound and require bleeding. The density meters must also be deactivated and zeroed when not in use, or the stagnant reading will be totalized with the good readings obtained while pumping.

The chlorination control is on manual flow control and the computer is just for monitoring or alarming. The compound loop control with the chlorine residual analyzer, although accurate, causes a sine wave type of control due to the lag time. A steady control is being accomplished with the manual control using flow, analyzer readings and experience as criteria. The problem of overfeeding at high flows and underfeeding at low flows is being studied so that an automated control procedure can be designed to compensate for these deficiencies if effluent chlorination is to be continued.

Before any process control can be implemented through digital techniques, the system must be thoroughly understood and field evaluated. TCB is still in that stage for many processes. The data-logging system is about four years old but the control capabilities have only been in existence a little less than two years.

This is brought about in part by another problem which affects the timeliness of all of the control programs—the availability of personnel to work on the system. TCB does not employ a full-time programmer. The programming functions are provided by personnel who have other primary duties. The size of the sanitary district has not warranted a fulltime employee as yet, even though one could find a great deal of backlogged work.

Future applications include continued emphasis on energy consumption, the installation and data analysis of an on-line ammonia monitor, inventory control and most of all the effective application of the existing instrumentation and control package including the process simulation.

MAINTENANCE

The atmosphere at the treatment plant can affect the digital control equipment. The high moisture and the hydrogen sulfide can attack solid state circuity and relays or contacts. These are constantly being cleaned and replaced. A controlled environment was needed for the computer system; this added significantly to the costs.

The reliability of the electric utility has also provided some headaches. The mass-memory disk drive is extremely sensitive to power fluctuations and even some minor voltage changes (or brown-outs) have wiped out the entire disk. The system must then be rebuilt from backup system tapes and hopefully backup data tapes so that up-to-date information has not been lost. Processes under computer control will also time-out and switch back to manual or local control.

At TCB there are two incoming power feeds and an emergency generator that starts automatically upon power failure and allows for the feeds to be switched. An uninterruptible power supply (UPS) has recently been purchased which filters out line changes, provides 15-60 min of battery supplied power for switching or blackouts and also has a slow slew frequency make up for the unstable generator source. This also has increased the capital costs, but has greatly improved the reliability of the entire system, especially when 6-12 fluctuations can occur during a spring electrical storm.

The maintenance of the hardware is also a prime consideration. It costs TCB an engineer's annual wages for a maintenance contract on the entire system. The coverage has been excellent and only one extended down period has been incurred since the system startup in 1978.

Other considerations for control are the field requirements. The atmospheric effects have been mentioned. Motorized gates and valves have also been discussed; and they require constant electrical and mechanical preventive maintenance. Ice has damaged several different models due to moisture in the controls or gears.

Large panels full of relays and 24-V power supplies are also needed if remote starting and switching capabilites are going to be provided. These also must be maintained, and good as-built terminal-to-terminal drawings are a must.

SENSORS

The real key to any process control or monitoring system is the sensors. The sensors are the weakest link in the control loop. Sensors can be affected by a variety of factors. They are the primary reasons many systems fail to prove useful.

The first problem is to specify sensors that work. A large plant can require that all prospective bidders pass a qualification field testing period and then verify it again when they are awarded the bid. Smaller plants must depend on the literature and examples set by others. An in-depth study should be made before accepting any sensor or specification.

Any sensor will fail if it is not maintained and calibrated regularly. TCB had dissolved oxygen meters that required constant cleaning and, while investigating alternative brands, contacts were made which showed that an improved understanding of the probes and modifying the methods of maintenance could make the existing type of sensor one of the most reliable. The difference in the membrane thickness between the field and laboratory units causes a slower response time, and when the cycling of the pumps causes a 1- to 2-ppm deviation in actual DO levels, the two meters will never read the same.

Similar problems have been encountered with suspended solids analyzers. All of the concentrated efforts were spent on the low-level sensors in the final effluent, where subsequent communications with other plants found that these were the least reliable and that the high-level meters used for mixed liquor or sludge analysis were the most beneficial. Part of the problem again is having the manpower to dedicate to the problem until a conclusion can be made. In these hard economic times, many plants are saving costs by keeping payroll to a minimum.

Many different types of flow meters are being used. Mechanical floats on a flume are dependable but do not always handle large fluctuations in flow well. However, the use of a flume assures that a reading can always be obtained; this is not the case with a mag-meter, for instance. Differential pressure transmitters for air flow are accurate but the tubing must be specified so that it does not deteriorate and crack. Measuring chlorine gas flow proved to be almost impossible due to its corrosive nature; therefore, indirect techniques are used. The vortex gas flow meters have only required minor maintenance and the thermocouples have proven to be almost maintenance free.

The computer itself, although limited by the quality of information provided, can augment that information itself. It can enhance the quality of the measurements through buffering, compensating for noise, analyzing the signal for reasonableness and even switching back to manual if necessary.

SUMMARY

There are many benefits to be realized. Tangible benefits such as cost savings due to increased efficiency are obtainable. Savings due to manpower are also real. The plant at TCB has more than doubled in size and complexity, whereas the personnel have only increased by about 10%.

The reports can be generated in minutes, whereas it used to take days. The information in the reports is always current, and this allows the operator to make better decisions and become a more skilled operator. This not only improves the quality and consistency of the effluent quality but also provides a certain amount of job satisfaction.

It is very important that all personnel be trained in the manual procedures and calculations so that they can perform in the event of a system failure. The system must also be programmed to remain flexible due to the constant changes in technology and requirements, and must be programmed with the operator in mind so that it is not so complex that only an engineer or programmer can understand it.

The complexity of many of the processes required to meet today's environmental restraints almost dictate that some sort of exact process control and monitoring systems be integrated into the treatment works. A computerized system is one way to satisfy those needs, and it can work for you if it is properly designed, installed, programmed and maintained.

CHAPTER 7

THE COMPUTER AS A TOOL FOR ACTIVATED SLUDGE

Robert G. Skrentner
 EMA, Inc.
 St. Paul, Minnesota

INTRODUCTION

Purpose

[Many activated sludge plants have computers. The application of these computers may range from monitoring to supervisory setpoint control to analog replacement control to various levels of process optimization. This chapter will describe the use of the computer for activated sludge control. It will present a systematic approach to move from basic control to optimized control. Optimization is achieved when effluent goals are consistently met while costs are minimized./

Approach

It is a rare occurrence to see an activated sludge plant start up and operate exactly as had been anticipated. Actual flow or loading characteristics may be different from those assumed during design. Individual processes or components may perform better or worse than expected. Plant personnel may have different skill levels than originally envisioned.

Most plants evolve over several years subsequent to startup. The operational strategy may change. Equipment and piping may be added, deleted or modified. Personnel will gain familiarity with the detailed plant operation and maintenance. In addition, every plant is unique in terms of both the process equipment and the organizational structure. Every plant has its own unique problems and potential for improvement. For these reasons, this paper will describe some experiences of other plants. Common to all of these plants is a computer-based control system, startup and operational difficulties and a commitment to success.

The Word "Tool"

One major advantage of a computer system is that it can be changed to follow the evolution in plant operation with little or no capital cost. Because of the flexibility of computer control systems, there is no need to expect the computer to do everything on startup. Designs can be simplified to provide very basic control. As more is learned about the process characteristics, the computer can be modified accordingly.

Computer control does not define a unique control method. It only defines a form of automation. Process control is not hardware, be it analog panels or computers. Process control is operations. People are responsible for operations and they use whatever tools are available to help them operate. Knowledge and commitment are required to assure success with any form of automation [1]. The key to success in optimizing the activated sludge process is not the computer, it is the use of the computer and the associated instrumentation as a tool by knowledgeable and committed personnel.

THE REALITIES OF ACTIVATED SLUDGE CONTROL

Activated sludge is an interesting process. It sounds easy: simply mix the incoming waste with a biological population, give plenty of fresh air, then allow the mixture to quietly settle. The process has been around for over fifty years. Its characteristics have been studied over and over. One would think that there should be no problem in optimizing the process.

Process Dynamics

To consistently meet effluent goals, the response characteristics of the process must be known. These characteristics include both the steady-state response and the transient response. To define these response characteristics is much more difficult than the cost accounting portion of optimization. Plants rarely achieve steady-state conditions. The flow and loading to the plant are constantly changing throughout the day and throughout the year.

Even if steady-state influent conditions could be approximated, it is difficult to change one variable, such as mixed liquor dissolved oxygen concentration, and observe the effect on the process. This is due to the process being relatively slow to react to change. A period of hours or days may be needed to observe the impact of the operational change.

Transient response testing must be limited to avoid causing major process upsets. Normal transient response can be observed by noting changes in loadings throughout the day. The impact of the change on the process is difficult to observe for the same reason cited previously.

Process Problems

Plant-scale testing of steady-state and transient behavior assumes that the process parameters can be accurately measured. It further assumes that some degree of plant stability has been reached and that the process has been demonstrated to perform adequately if not optimally. In many plants, this is not the case. Flow and solids measuring may be unreliable or nonexistent. There may be significant hydraulic or solids imbalances among the aeration tanks or final clarifiers. The plant may be over- or under-designed [2]. There may be significant operational flexibilty limitations [3].

Process Control in Practice

Many plants are happy to get by with an occasional discharge limitation violation. True optimization is not even considered. Fortunately, the activated sludge process is a "forgiving" process. For example, a dissolved oxygen concentration of 2 mg/l might be desired. However, it may vary from 1 to 5 mg/l or more during the day. The same is true for the mixed liquor solids and the total mass within the system. Most plants can be operated within fairly wide limits without detrimentally affecting the process.

The above is frequently cited as an excuse to forego any type of automation and especially computerized automation. "Why do I need a computer control system when my operators can make one or two adjustments per day to the return or waste rate or the air addition rate?" One plant found that closer control of dissolved oxygen concentration could save $20,000-60,000/yr [4]. A small tertiary plant abandoned its chlorine residual analyzer (and the resulting operation and maintenance costs) when it was found that the predictive control model could maintain residual within very tight limits. Prior to that time, the chlorine savings were being offset by the expense of the analyzer.

GOAL SETTING

At one small two-stage activated sludge treatment plant, the computer control system contained 47 different control strategies. These included bar screen control, primary sludge withdrawal, air blower and dissolved oxygen control, clarifier sludge withdrawal, return and waste pumping based either on operator setpoints or various predictive models, tertiary filter control and chlorination, to name a few. When the plant was commissioned, only two of the control strategies could be used. Instruments, field control panels, field wiring and field control devices had not been thoroughly checked out prior to startup. Three years later, 36 control strategies are being used. Additional strategies are being started up as the plant field equipment is corrected or replaced.

The plant manager was committed to making the plant as efficient as possible. He prioritized areas requiring improvement and systematically solved many of his problems. He used many of the success steps in the action plan which follows.

Existing Plants

Those considering upgrades to the control system for an existing plant or process should have some familiarity with the existing operational constraints. This makes the task of upgrading somewhat easier. In an action plan, you must decide what to modify, when to modify it, and how extensive the modifications are to be.

In one large plant, it was decided to replace the existing central digital control system with a distributed computer control system. Very few modifications were made to existing instrumentation. The action plan and the distributed control concept were based on several years of operating experience. It considered the operator's skill levels and the problems associated with a major change in control philosophy and operational transition period. In another plant, older instruments and control devices are being replaced to provide more reliable control and to interface with an existing computer system. No major operational changes are anticipated.

New Plants

For new plants or those undergoing major process additions, it is more difficult to establish an action plan. Since there is no operational experience base, it is difficult to know what to expect. It is critical that new plants anticipate potential problems and develop an action plan which has built-in problem prevention and contingent actions [5].

AN ACTION PLAN

There are three major phases in moving from basic control to optimized control. At the end of the first phase, the characteristics of the process should be measurable. These characteristics include flow and flow splitting, solids and solids splitting, dissolved oxygen, levels, pressures and other process parameters. The characteristics chould be measurable by on-line instrumentation or by a defined program of manual sampling and testing. Confidence is gained in basic control.

At the end of the second phase, the plant should start to see improvement in energy utilization, less reliance on the operator to make routine control decisions, and a more stable operation.

The third phase is a result of much experimentation and analysis to define the response characteristics of the process under different loading conditions and to formulate predictive models to compensate for these changes.

These phases do not necessarily have clear points of distinction. For this reason, each major subprocess is discussed in terms of moving from basic control to optimized control. In attempting to optimize control, solving one problem may lead to another. For example, tighter air addition control may point out weaknesses in blower control. Some of the cases presented show how these additional problems were solved.

Flow and Solids Distribution

There seems to be a variation of Murphy's law which applies to solids and flow distribution within a plant. This could be worded as follows:

> If the flow splits equally among units, the solids won't.

A corollary to this law might be:

> If solids loading is equalized, hydraulic overloading will occur.

The easiest way to verify this law is to control all flows equally to the process units and watch what happens to the solids.

Figures 1 and 2 show some typical plant layouts for both aeration tanks and final clarifiers. Figure 1 shows examples of controllable distribution

92 ACTIVATED SLUDGE PROCESS CONTROL

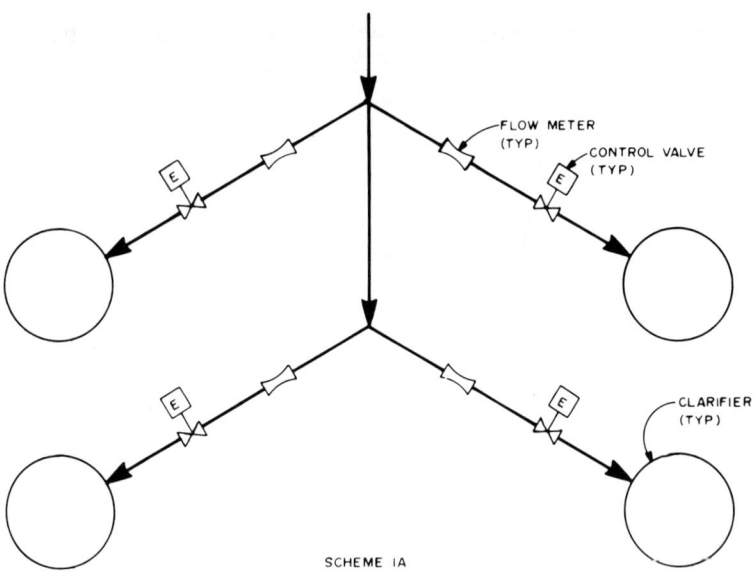

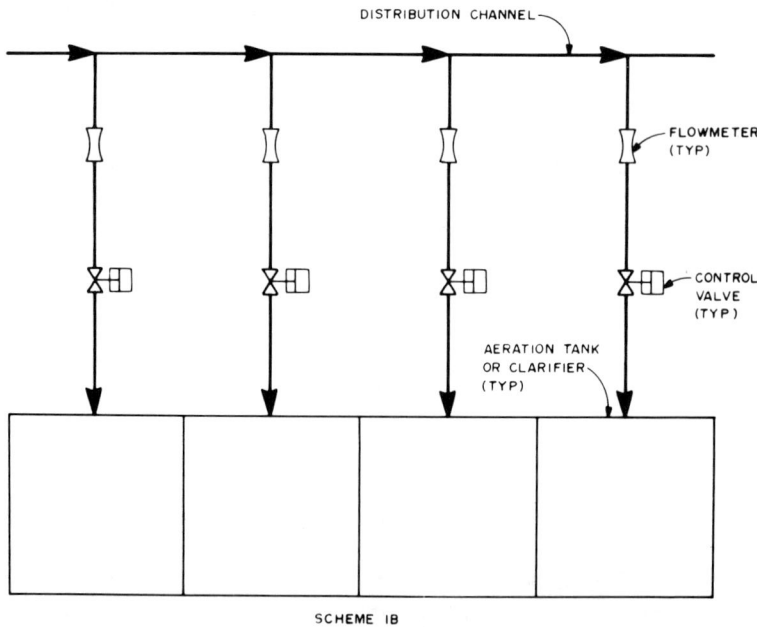

Figure 1. Controllable flow distribution.

THE COMPUTER AS A TOOL 93

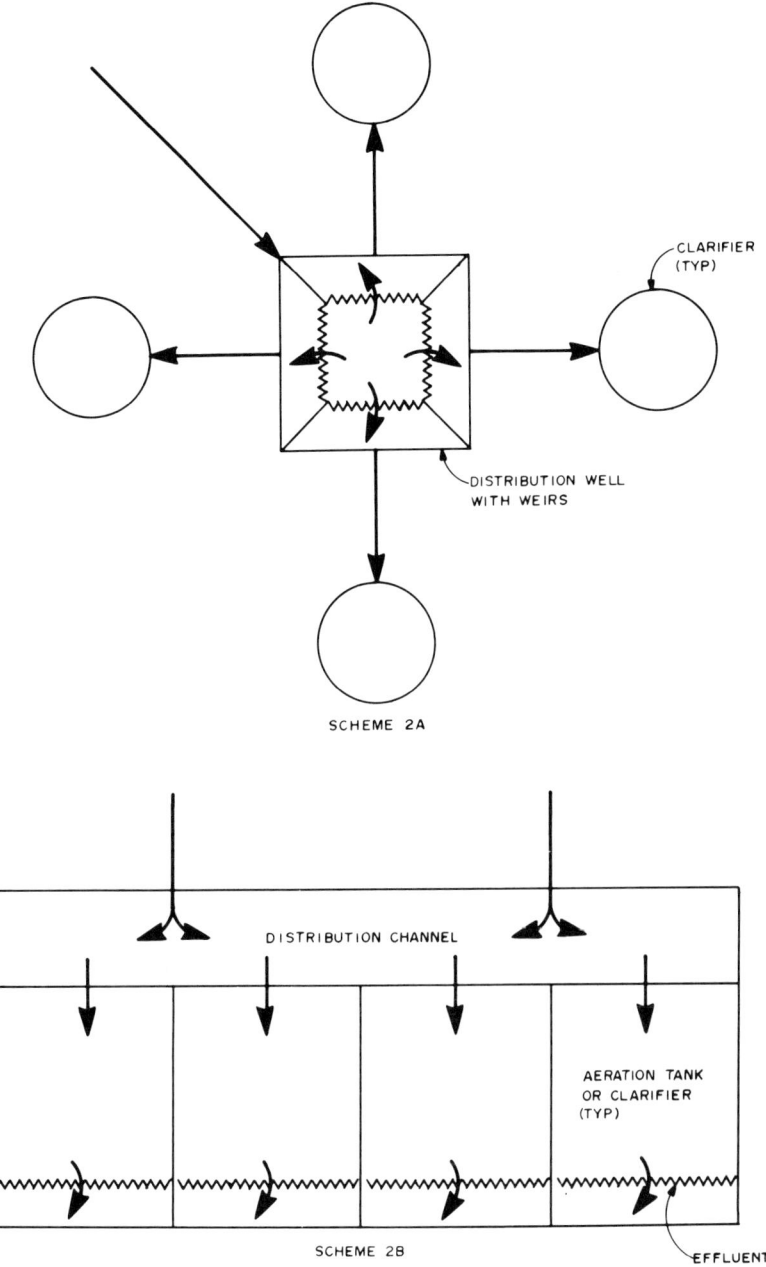

Figure 2. Uncontrollable flow distribution.

schemes and Figure 2 shows examples of uncontrollable distribution schemes.

Flow Measurement and Control

The first step to quantify flow and solids splitting characteristics is to accurately measure the flows. The City of Detroit [6] has over 50 magmeters in the secondary process. In an attempt to computer control flow splitting, the accuracy and reliability of these meters was questioned. An extensive evaluation program found many problems with the installation of the meters. Many of these problems were corrected. The computer was modified to compare valve position to flow and to detect flow meters which might be out of calibration.

At the Sioux City plant [7], the pitot tube magmeters were found to be totally unreliable for flow splitting control. The meters were abandoned in place. Valve position indicators were installed on the control valves and the computer was modified to control valve position to split flows.

Figures 3 and 4 show the computer control schemes for the two plants. Backup controls are not shown. To account for uneven solids distribution, the control scheme included a ratio station. Based on the solids buildup in the final clarifiers, the operator could adjust the ratio to each tank to more closely balance solids. The flow and valve position controllers have adjustable limits to prevent hydraulic overloading.

Another large plant found that flow was uncontrollable during low flow periods. The influent meters and valves were sized for peak flow conditions. (This is the case in many plants.) Since flow and solids distribution were not critical during low flow periods, the control was modified to turn off the flow control when the total flow fell below an operator-set minimum value. This stopped the excessive hunting of the valves.

Those who have weir splitting schemes, as shown in Figure 2, may be thankful to have avoided some of the preceding problems. However, with the possible exception of scheme 2B, it is unlikely that the solids are equally distributed.

One plant with final clarifiers arranged similar to those in Figure 2A found that the clarifiers nearest the mixed liquor influent lines on the distribution header were consistently overloaded and that sludge could not be removed fast enough without creating a "hole" in the blanket. The clarifiers were equipped with isolation sluice gates for maintenance purposes. It became necessary for the operators to adjust the gates approximately once per shift to attempt to balance flows. After some experimentation, three gate position settings were found to be required,

THE COMPUTER AS A TOOL 95

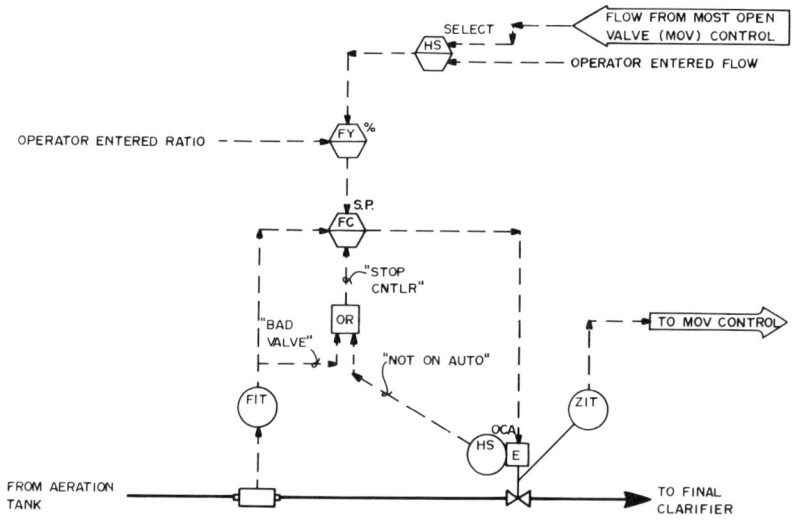

Figure 3. Basic flow control with ratio station.

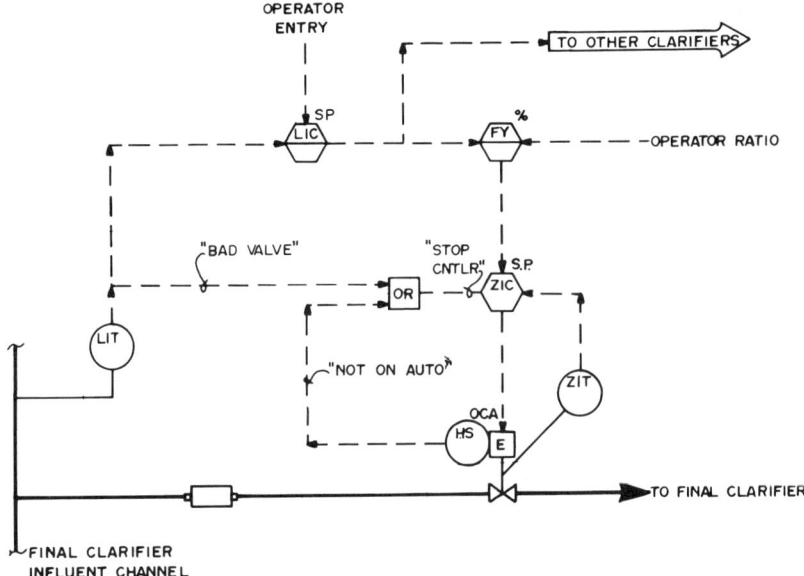

Figure 4. Cascaded level to position control with ratio station.

one each for high, medium and low flows. Since 96 gates were involved, this kept one operator busy for a good part of the shift. The computer is being modified to move the gates. It does this in sets of 4 gates at a time to avoid overloading to the plant power distribution system.

Flow Transients

One primary plant had constant-speed pumps in the main lift station. When secondary treatment was added, the hydraulic transients created by these pumps adversely affected the final clarifiers. Modulating valves are being added to the pump discharge to reduce the sudden flow changes. At another plant, upstream interceptor lift stations were creating a similar problem. This was solved by retuning the pump controls to provide slower response.

Both plants detected these problems while attempting to provide better flow control within the secondary process. The sudden flow changes had required very fast control response. This should be avoided in the activated sludge process.

Solids Measurement and Control

The preceding flow control modifications were performed without on-line solids instrumentation. Lab or operator tests for solids concentrations and sticking of clarifiers for blanket level were used.

The next step in flow and solids distribution control *could* be the installation of on-line solids analyzers and blanket level probes. This type of instrumentation has been proven to work if properly installed and *maintained* [7,8]. If you are having trouble maintaining your flow meters, it would be better to forget about this type of instrumentation and use manual methods [9].

One plant installed the mixed liquor suspended solids analyzers in sample sinks. This allowed the operator to very quickly perform custodial maintenance on the probe on a daily basis. He simply lifted it out of the sink and washed it off. The probe was not of the self-cleaning type. In another instance, probes were relocated upstream of the return activated sludge pumps. This stopped the seals from leaking and reliable measurements were obtained.

Solids measuring instrumentation should *not* be acquired under competitive bidding. It is recommended that a 60- to 90-day test be conducted on these instruments for the particular application. Test as many different types as possible and pick the one that best matches your operation and maintenance capabilities. Only after you are satisfied that the unit will work should it be acquired. It will be worth the fight with the purchasing department or the EPA to use sole source procurement.

Whether or not you decide to acquire this type of instrumentation, the control strategies can be modified to account for solids. Figure 5 shows a control scheme that accounts for solids loadings. Since the process is relatively slow to respond to solids changes, the calculations may be based on on-line instrumentation or periodic (every 4 to 8 hr) manual entries. Notice that various operator override capabilities are provided to allow for solids sensors being out of service. This strategy is normally used for return activated sludge control. It could be used for influent control.

By careful control of the flow and solids distribution, the process is better able to respond to changes in influent characteristics. This will result in more consistent performance in meeting effluent goals.

Sludge Handling and Distribution

Removing sludge from the final clarifiers usually involves pumping. Typical sludge withdrawal schemes are shown in Figures 6 and 7. Since pumping requires energy, this is a good spot to look for areas to reduce operating costs. However, prior to reducing costs, other process control considerations must be addressed. The sludge is withdrawn from the clarifier, most of it is returned to the aeration tank, and a small amount will be wasted. As in the case of flow and solids distribution, the first task to achieve the preceding is to implement basic flow control.

Clarifier Sludge Withdrawal Control

Depending on the plant operating staff and the piping configuration, implementing basic flow control may or may not be a relatively easy task. Two potential problems immediately arise:

1. What flowrate should the operator set for withdrawal?
2. What is the impact of flowrate changes on downstream processes?

The answer to the first question depends on how the process is to be operated. Some plants set a constant return sludge rate and allow the blanket level in the clarifier to vary. During high loading periods, the clarifiers store sludge. Other plants give priority to the blanket level and vary the amount of sludge withdrawn from the clarifier to keep the blankets as low as possible. In this case, the return to the aeration tanks is variable. Still other plants set a return rate in proportion to the influent flow in an attempt to maintain a constant food to microorganism ratio.

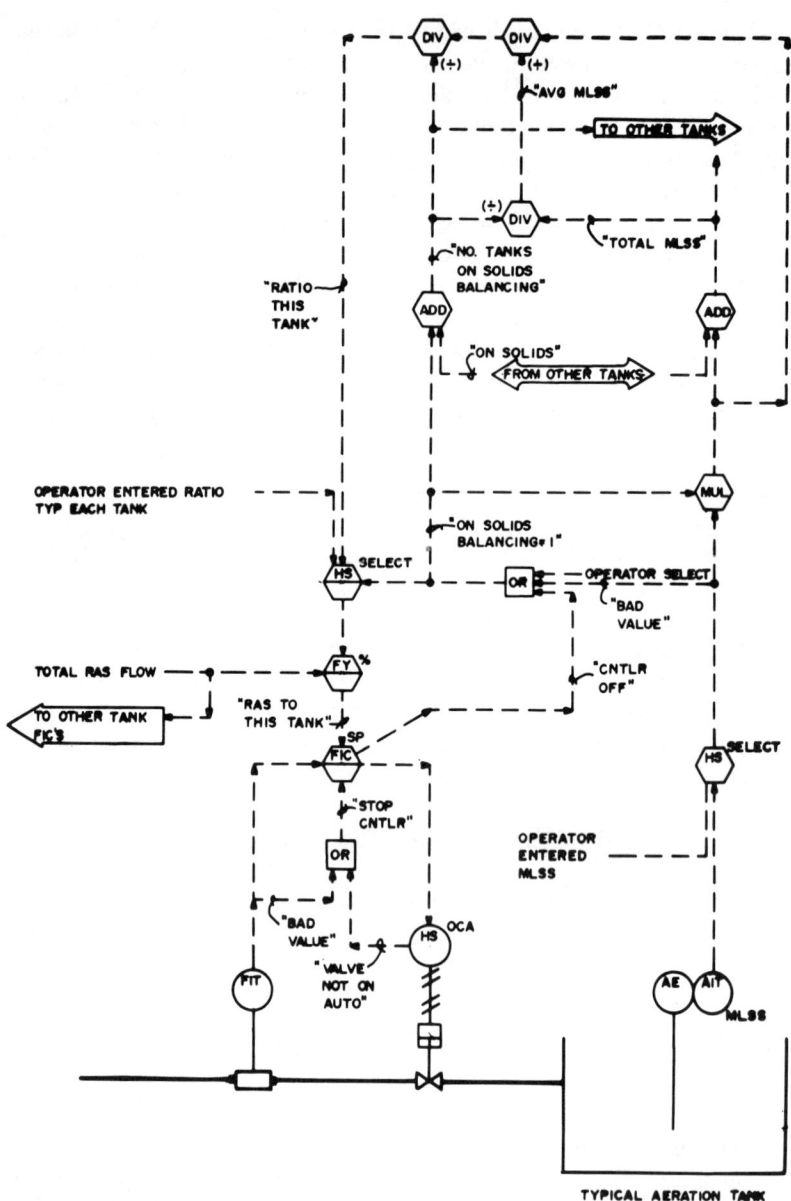

Figure 5. Mixed liquor solids balancing control.

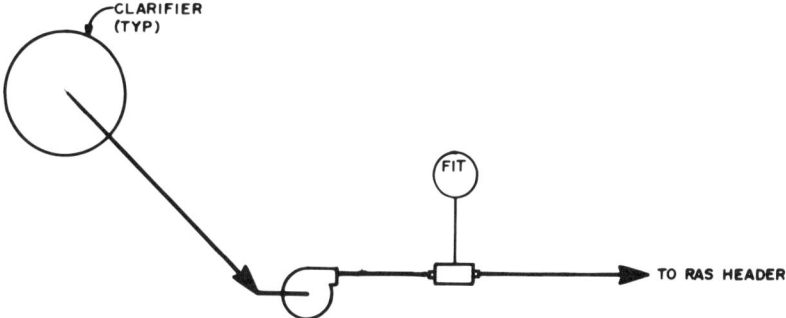

Figure 6. Pumped clarifier sludge withdrawal.

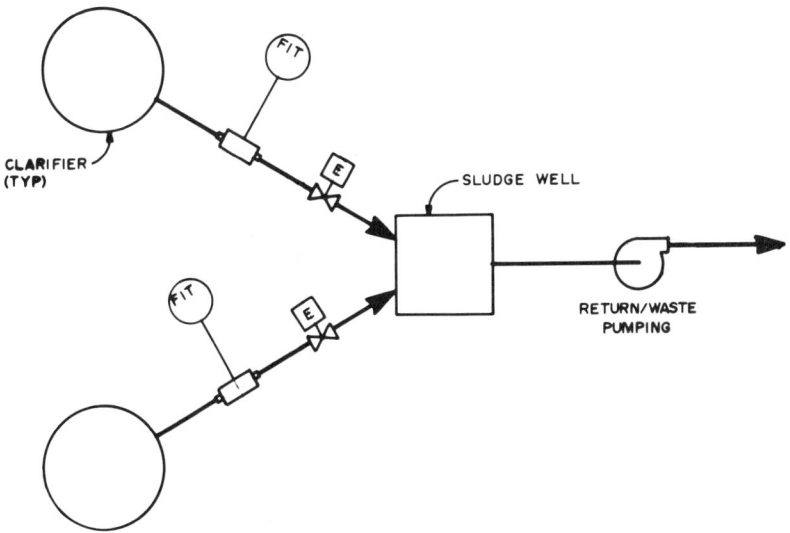

Figure 7. Gravity clarifier sludge withdrawal.

From one of these three basic process control strategies (or other strategies) a withdrawal rate can be determined. Figure 8 shows a control strategy for this case.

To make the job easier, the operator enters the total desired return and waste sludge. The control strategy compensates for the number of

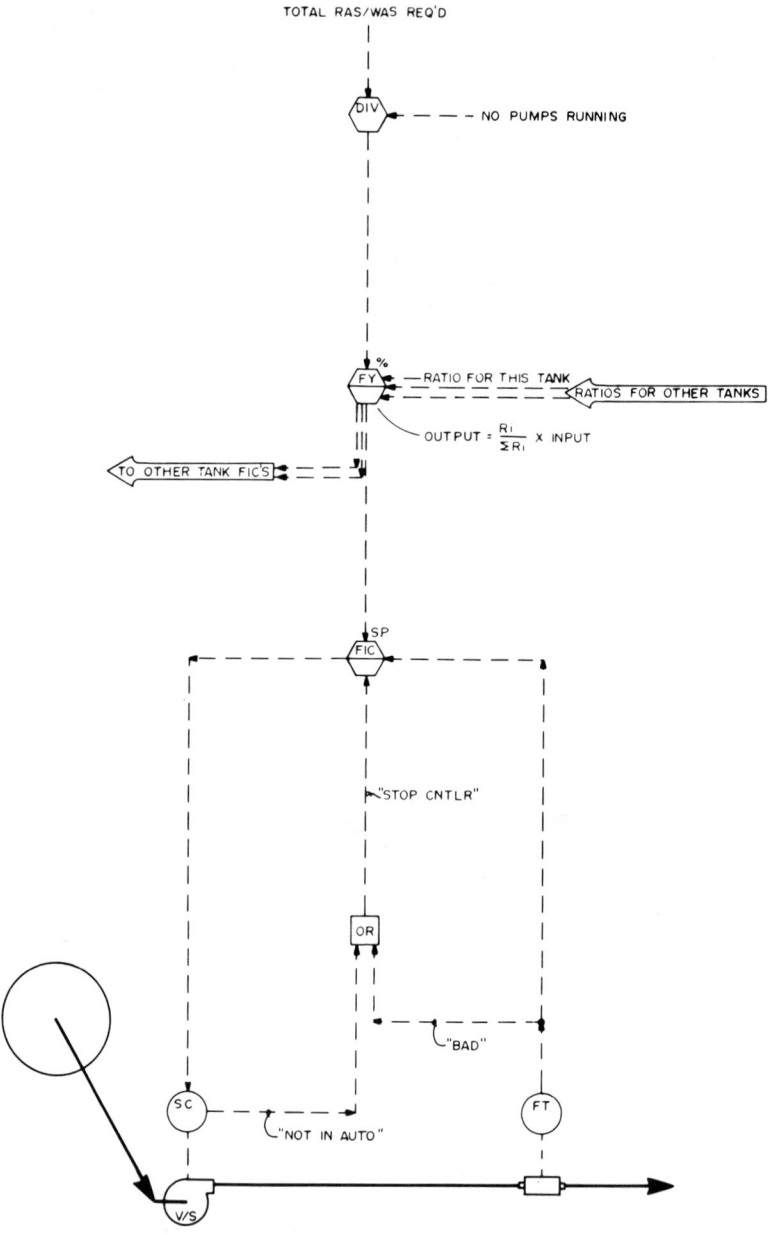

Figure 8. Clarifier sludge withdrawal control.

clarifiers in service. It also contains a ratio to allow the operator to compensate for any difference in sludge blanket level among clarifiers.

The answer to the second question depends on your process. In one plant with a layout similar to that in Figure 6, telescoping valves in the return sludge well were oversized. Control was so sensitive that plant operators could not properly set withdrawal rates. The valves were lowered fully and the pumps were used to control the withdrawal rate. Fortunately, unequal sludge buildup in the clarifiers was not a problem.

In two larger plants, the return sludge piping hydraulic capacity was limited. If an operator wanted to increase withdrawal rate from one clarifier, he would have to decrease the rate from other clarifiers. In these two cases, the computer controls were modified to reflect required changes in the operational philosophy. For the telescoping valve case, the control system could not overcome the valve sizing problem. For the limited capacity case, the flow decrease required was calculated and split among the remaining tanks by the computer.

Once basic flow control has been implemented, more complex control schemes can be added. These may use blanket level probes, solids analyzers or both. Common withdrawal control strategies that have been used successfully include:

1. **Automatic increase or decrease in clarifier withdrawal rate to compensate for sludge blanket level high or low alarms:** Normally, this has been limited to a fixed percentage increase or decrease in withdrawal rate. If normal blanket levels are not reached within a fixed time period, an alarm is given to the operator. The percentage change and time delay are based on experimentation with the process.
2. **Equalizing withdrawal solids:** As an alternative to blanket probes, one plant adjusts the clarifier sludge withdrawal rate to maintain equal solids concentration out of each tank. This control assumes that the sludge compaction rate reflects blanket levels.

Both of the above build on the simple withdrawal sludge pumping control. As an alternative to the operator-entered ratios, the computer is used to calculate ratio changes. This control scheme is shown in Figure 9.

Return Sludge Calculations

There are numerous models available for predicting return sludge rates [10,11]. These can range from simple flow proportioning to various predictive models based on biological oxygen demand (BOD), total organic carbon (TOC) or other measures of the incoming waste strength.

For computer-controlled systems, it is the practice of EMA, Inc. to include three to five different models to calculate return sludge rates. All

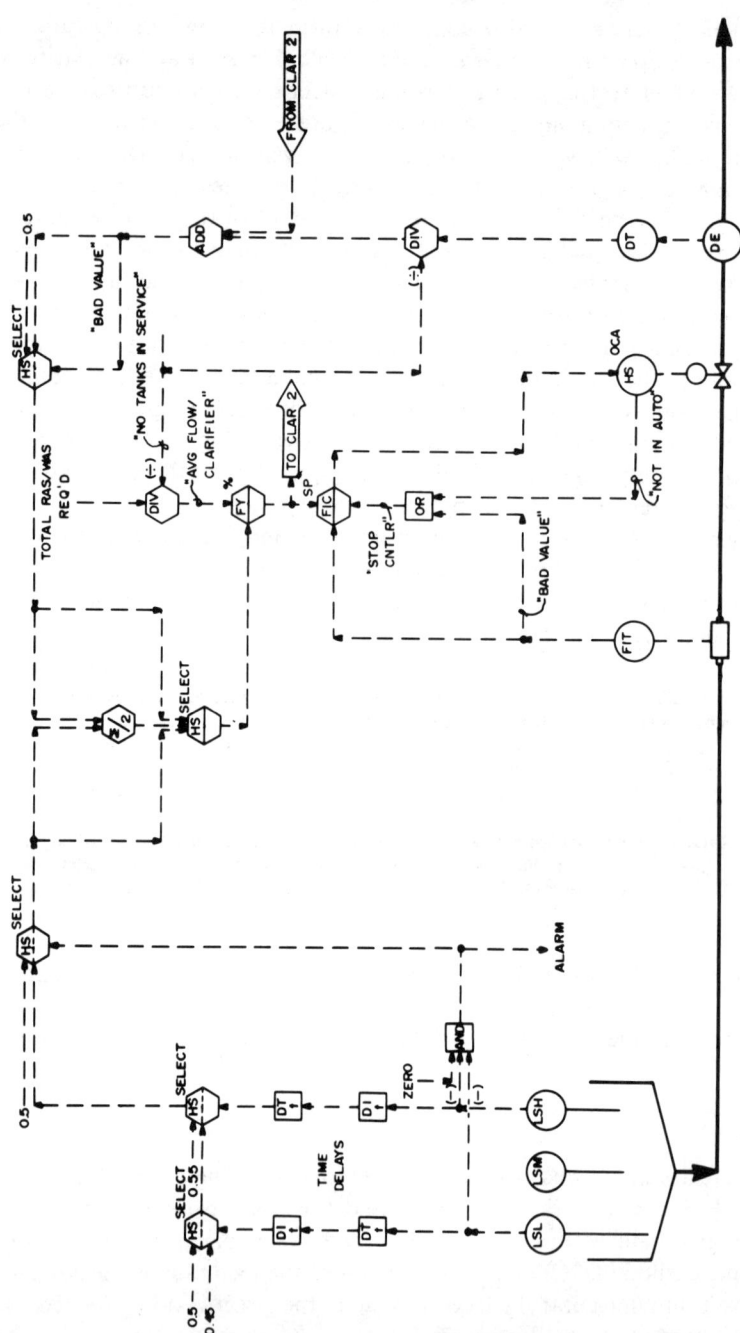

Figure 9. Clarifier sludge withdrawal with automatic blanket override and density compensation control.

models continuously calculate and display results. The operator selects the desired model. In general, these have not been used for one or more of the following reasons:

1. Most plants get by with simple flow proportioning or operator-entered rates.
2. The plants have not yet mastered simple flow and solids control.
3. The analytical instrumentation is not maintained.
4. Other plant areas are getting priority.
5. Experimentation requires time and personnel which are not available.
6. There is too great a risk for process upsets with full-scale testing.

There is no need to use more advanced calculations if the basics will work. There is also no need to defer trying more advanced ideas to further optimize performance. One advantage of having several models is that space is reserved in the computer for future enhancements or replacement of existing models.

Return Sludge Control

Return sludge may be introduced into the influent conduit or directly into the aeration tanks. When introduced directly into the aeration tank, several advanced control strategies have been used. In several plants, flow ratios are adjusted to help equalize mixed liquor solids. This control strategy is shown in Figure 5. In another plant, a similar approach is used but the operator adjusts the ratios based on periodic solids tests.

Waste Sludge Control

Wasting may be from the mixed liquor or return sludge lines or both. Waste rates normally are set to maintain a fixed level of solids in the system. In many plants, the total pounds to be wasted is calculated on a daily basis. From the latest solids data, the daily waste flowrate can be established. Although it is likely that solids concentration will change somewhat during the day, this mode of wasting usually is sufficient.

As with most continous processes, it is beneficial to maintain steady-state conditions. A relatively constant waste rate over the day would be desirable. Some plants cannot waste continously, especially for low flowrates; intermittent wasting is required. This may be a batch operation, in which the solids are removed continously for a part of the day, or a cyclical operation, in which wasting is performed periodically throughout the day. In either case, downstream processes may be adversely affected. Smaller pumps or control valves for low flowrates may be required.

Enhancements to the basic wasting control may include:

1. **Automatic calculations to compute the pounds to be wasted:** This may be based on lab- or operator-entered solids data or from on-line instruments. Sludge retention time (SRT) or other models can be used to calculate the waste amount.
2. **Constant mass withdrawal:** Based on the total pounds to be wasted and the current return sludge (or mixed liquor) solids, a daily average flow can be computed. The controls can perform continous, batch or periodic wasting as required. Periodic updates based on the current solids can ensure that the desired total waste is met. A control strategy is shown in Figure 10. This control enhancement may also help optimize downstream sludge treatment processes.

Energy Considerations in Sludge Handling

One plant uses sludge wells and pumps similar to Figure 6 for clarifier sludge withdrawal. During low flow periods, the sludge well level can be kept high since sufficient gradient is available to remove sludge. At higher flows, the level must be lowered. Dynamic sludge well level setpoint adjustment is used to maintain the highest possible suction head on the pumps. This will save at least $10,000/yr in power costs.

Another plant pumps sludge into a common header for return to the aeration tanks. Flow is split to the tanks using cascaded control based on header pressure. Under low flows, a header pressure of 6–8 psi is required. Under high flows, 12 psi is required. By dynamically adjusting the header pressure setpoint, pump power requirements are reduced.

The preceding cases save a relatively small amount of energy when compared to aeration power costs. They serve to demonstrate areas where costs savings can be realized.

Dissolved Oxygen Control—Air Systems

In air activated sludge systems there are usually two major control loops—the air addition and blower output.

Dissolved Oxygen Control

The first task in dissolved oxygen (DO) control is to find a probe you are happy with. Acquisition of DO probes should be the same as that of solids probes. Testing and evaluation are required. All probes should be maintained weekly even if the testing shows longer periods between maintenance to be possible.

THE COMPUTER AS A TOOL 105

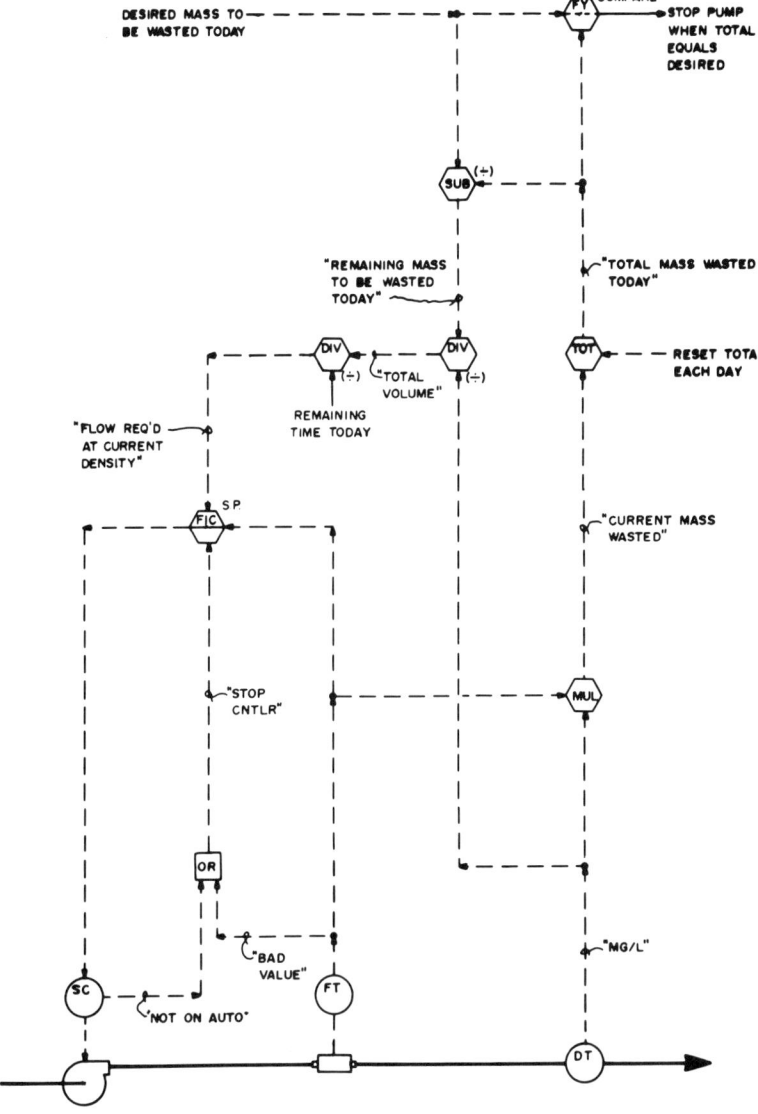

Figure 10. Constant mass waste sludge control.

DO can be difficult to control because of the relatively long process time constants. The effect of change in air flow may not be detected by the DO probe for 10-20 min or more. The same is true if the air flow is held constant while the influent flow changes.

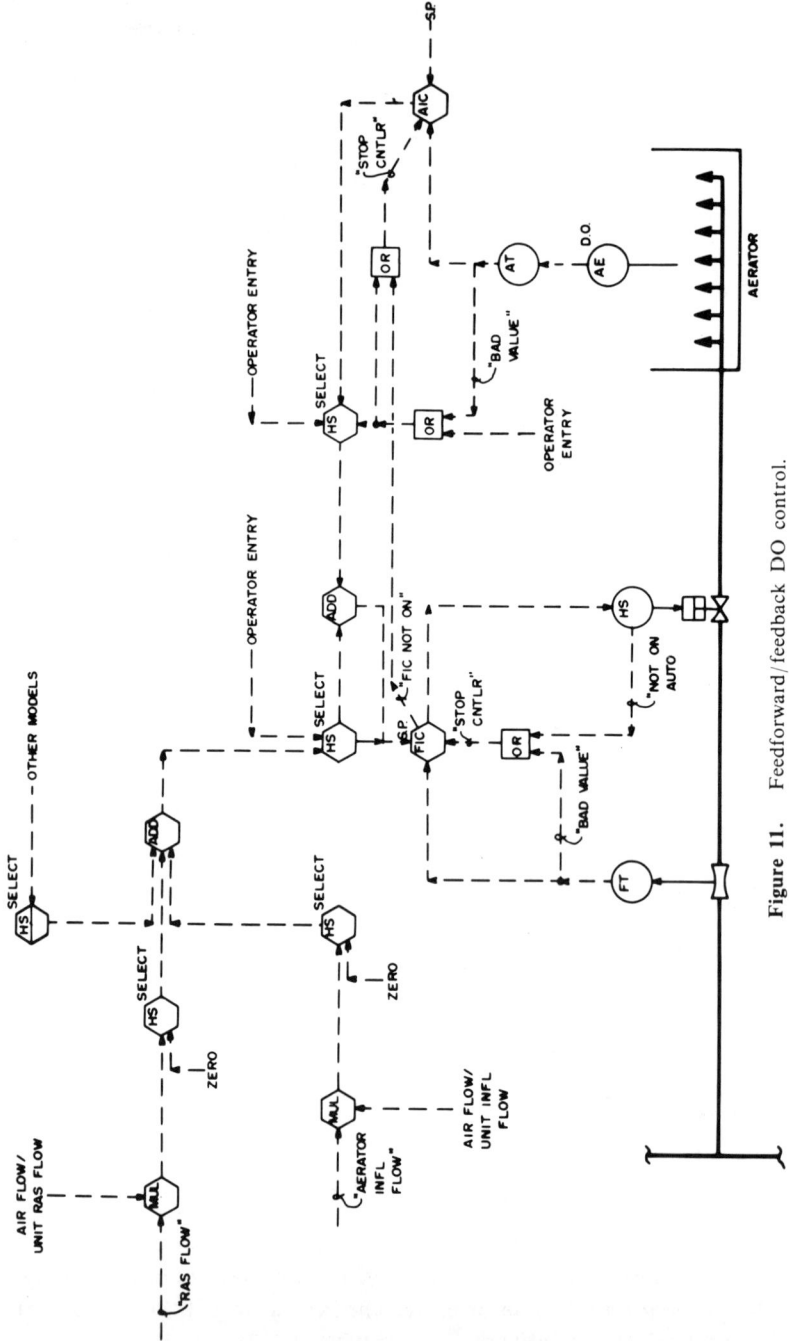

Figure 11. Feedforward/feedback DO control.

Neglecting the problem of decoupling blower control loops from the air addition loops for the moment, many plants have added feedforward control to more accurately control DO. Figure 11 shows a basic feedforward/feedback control scheme. The scheme allows the feedforward control to be based on several options. These may include influent flow, return sludge flow, models or combinations. Other enhancements to the basic feedforward/feedback loop could include:

1. feedforward based on influent plus return sludge flow
2. feedforward based on mass loading of influent, return sludge or both
3. feedforward based on total organic carbon loadings or respirometers or other instruments which provide a measure of potential oxygen demand

Each of the preceding enhancements is intended to optimize the air addition. When air flow can be adjusted to consistently produce the desired DO, blower control can then be optimized to minimize power consumption.

Blower Control

In the Piscataway plant [4], the computer controls were modified to provide fully automatic blower control including startup, shutdown and surge protection. This was reported as necessary to decouple the DO and blower loops. The airflow control is deactivated whenever blowers are brought on-line or taken off-line. The "stopping" of flow controllers is easy to implement with computers. It is much more difficult to perform with analog control.

In other plants, the blowers are started and stopped by an operator at the blowers. This can result in control loop windup for DO control. For example, if the air requirements are more than the on-line blowers can produce, the air valves will open fully in an attempt to provide the necessary air. When another blower is started, a large increase in airflow will result. Even with fully automatic control, this can occur because of the time required to bring another blower on-line.

In another plant, whenever the air header pressure falls below a predefined value, all airflow control is stopped. The computer decides which blower should be started and lights a light on the blower building control panel. It then ramps all on-line blowers to minimum output and waits for the operator to start the blower. When the blower is on-line, the header pressure control is activated. After a time delay to allow the header pressure to stabilize, the airflow control is restarted.

Figure 12 shows a possible control scheme to stop DO control

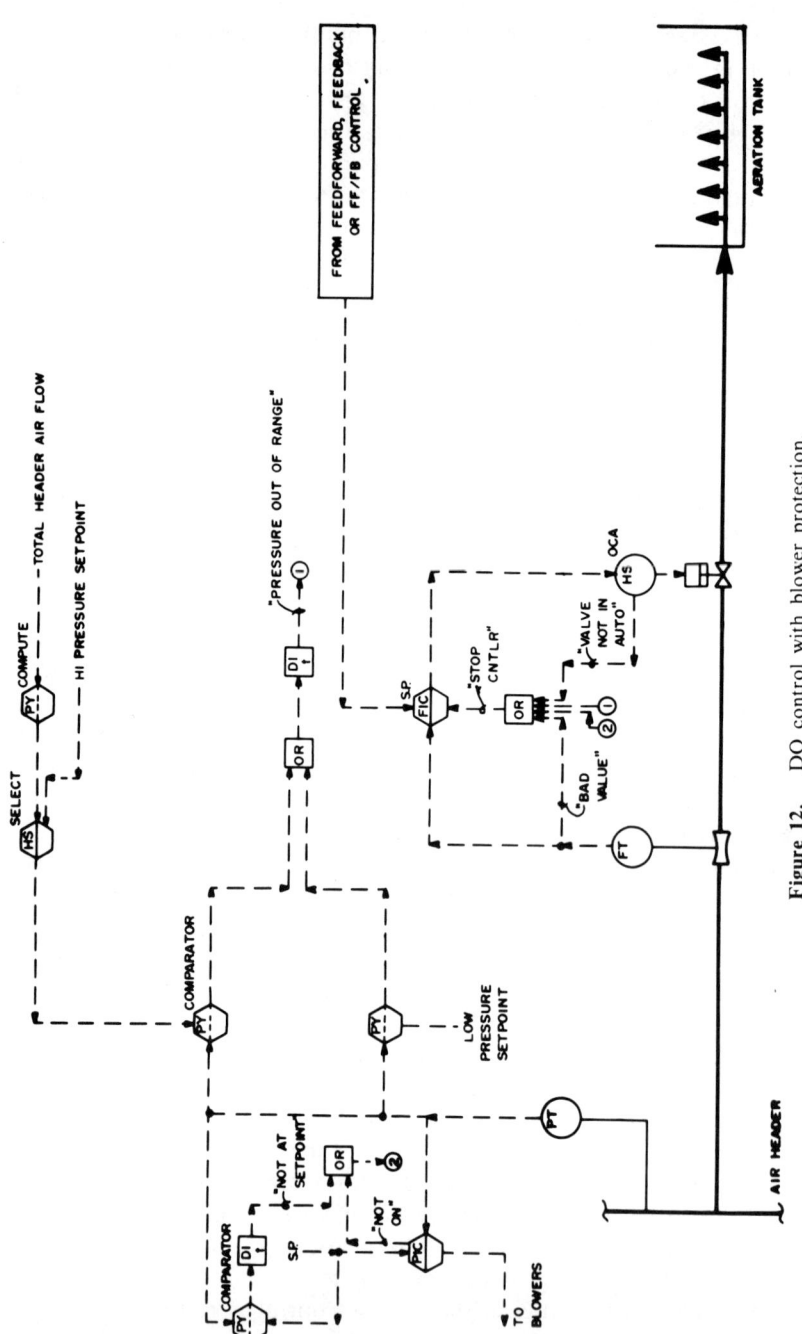

Figure 12. DO control with blower protection.

whenever the header pressure is not within normal operating range. This could help to prevent blowers from reaching a surge condition.

Dissolved Oxygen Control—Oxygen Systems

In oxygen activated sludge systems, the control usually maintains the first-stage pressure at a value sufficient to provide the required oxygen. Vent gas purity or DO probes may or may not be used to further refine control.

The largest potential for savings is in the oxygen generation. Monitoring of the oxygen plant can indicate trends in power consumption per pound of oxygen produced. It is also possible to provide some turndown control to reduce output during low oxygen demand periods.

Miscellaneous Controls

Other control loops in the activated sludge process may include pH, polymer, alum or pickle acid addition. Each of these loops can be optimized. Flow control with manual tests to monitor performance is the first step. Additional instruments can be added as the needs dictate to provide tighter control. Power and chemical use can be minimized.

SUMMARY

Some Observations

1. The staff at successful plants realize that EPA "ten-cent dollars" may not be the solution to their problems. Some are allocating a portion of their operating budget in order to get what they want instead of what the regulations may thrust upon them.

2. Staff are recognizing that "crystal ball procurement" will not lead to success. Rather than specify and acquire instrumentation under competitive bidding, plants are borrowing, leasing or buying instruments for testing and evaluation. Instruments proven to be cost-effective in the particular plant environment are then purchased on a sole source basis.

3. Less reliance is being placed on vendors. Plant personnel are becoming more knowledgeable in the area of instrumentation and controls.

4. Plant personnel are becoming more realistic in their expectations of what a computer control system can and cannot do. There is an awareness of what it takes in terms of time and resources to optimize the process.

Success Characteristics

The activated sludge process can be optimized. The computer can be a valuable tool in process optimization. In the cases presented, there exist several factors which lead to success. Some of these are discussed in the following sections.

Management Commitment

Management at these plants was willing to commit the time and resources required for:

1. analysis and experimentation
2. process modification
3. evaluation and procurement of proper instrumentation
4. proper maintenance
5. operator training

Computer-Based Control Systems

The computer-based control systems facilitated control modifications. As more was learned about the process, controls were modified to obtain the most cost-effective operational strategies.

Control Awareness

This reflects a plant staff awareness of the true meaning of automated control systems. It is not individual control loops, computer systems or instruments. Rather, it is a collection of people who utilize the tools available in the best practical manner.

REFERENCES

1. Knudsen, D.I. and R.C. Manross. "Computer Control of Wastewater Treatment Plants—It Takes Knowledge and Commitment," paper presented at the 51st Annual Conference, Water Pollution Control Federation, Anaheim, CA, October, 1978.
2. Vasicek, P.R. "Use of a Kinetic Study to Optimize the Activated Sludge Process," *J. Water Poll. Control Fed.* 54 (8) (1982) pp. 1176-1184.
3. Gray, A.C., Jr., et al. "Evaluation of Operation and Maintenance Factors Limiting Biological Wastewater Treatment Plant Performance," U.S. EPA, Report No 600/2-79-078 (Washington, DC: U.S. Government Printing Office, 1979).

4. Fertik, A., et al. "Computer Control of Air Nitrification," paper presented at the ISA Fall Conference, Philadelphia, PA, October 1982.
5. Kepner, H., and B.B. Tregoe, *The Rational Manager* (Princeton, NJ: Kepner-Tregoe, Inc., 1965) pp. 207-228.
6. DeLaura, T., et al. "Magmeter Evaluation Program for the Detroit Wastewater Treatment Plant," paper presented at the 1981 Water Pollution Control Federation Conference, Detroit, MI.
7. Graupmann, R.W., et al. "Start-up and Operating Experience with Computerized Control System in a 1.3 m 3/s (30 MGD) Wastewater Treatment Plant," *Water Sci. Technol.* (G.B.) 13:405-411 (1981).
8. Dedinsky, H. "What is Needed to Successfully Automate a Wastewater Treatment Plant," Milwaukee Metropolitan Sewerage District.
9. West, A.W. "Operational Control Procedures for the Activated Sludge Process," U.S. EPA Report No. 330/9-74-00ld (Washington, DC: U.S. Government Printing Office, 1974).
10. Keinath, T.K., and B.S. Cashion, "Control Strategies for the Activated Sludge Process," U.S. EPA Report No. 600/2-80-131 (Washington, DC: U.S. Government Printing Office, 1980).
11. Keinath, T.M. "Solids Inventory Control in the Activated Sludge Process," *Water Sci. Technol.* (G.B) 13:413 (1981).

CHAPTER 8

RELATING BOD₅ WITH ON-LINE OXYGEN UPTAKE RATE MEASUREMENTS USING AUTOMATIC RESPIROMETERS IN VIEW OF PROCESS MONITORING AND CONTROL

Dr. Normand Therien and Ferhat Ilhan
University of Sherbrooke
Sherbrooke, Quebec, Canada

INTRODUCTION

Activated sludge wastewater treatment plants can be sensitive to large disturbances in the flowrate and/or quality of the incoming wastewater. When this occurs the system receives a shock loading, resulting in lower treatment efficiency. The fundamental problem then lies in the determination of control policies which can overcome the effect of such disturbances. However, the success of implementing an adequate control action is highly dependent on monitoring certain parameters of the process. With the advent of reliable on-line sensors, systems can now be monitored and corrective actions taken rapidly without recourse to lengthy off-line laboratory analysis.

This chapter discusses conditions under which reliable estimates can be obtained for the five-day biological oxygen demand (BOD_5) of biodegradable wastewater using automatic respirometers.

RESPIROMETRIC DETERMINATION

The determination of BOD by respirometric methods using continuous recording of dissolved oxygen uptake is a common procedure, and several types of laboratory and on-line respirometers now exist. However, the oxygen uptake measurements obtained from such an apparatus are rarely usable directly as a means of precisely estimating the BOD of the liquor being sampled. This arises from the fact that the uptake rate is a function of both the metabolic state and concentration of the biomass in contact with the liquor, other abiotic environmental factors (temperature, pH, etc.) held constant. The assumption when conducting such a test that microorganisms are in the endogenous state is often incorrect, and invalid results may be obtained.

Biomass may also prove a difficult determination when computing the specific uptake rate. This is often alleviated by assuming biomass to be related to the volatile suspended solids (VSS) content of the liquor used. However, this assumption is often questionable. Usually a better estimate of biomass is obtained by relating it to the rate of its endogenous respiration. Another measure of biomass can be obtained by contacting it with a totally biodegradable substrate above the critical substrate/biomass concentration ratio. Under such conditions biomass limits the substrate assimilation process, and the oxygen uptake rate will provide an indirect measurement of active biomass. This concept has been used in this study in correlating the BOD_5 of liquors of varied nature with oxygen uptake rate measurements using a laboratory respirometer.

Background

The possibility of predicting the BOD_5 in influent wastewater has been demonstrated by Arthur and Hursta [1] using an automatic respirometer technique and relying only on the original microbial seed of the liquor. Due to the low initial biomass concentration, long periods of time were necessary to obtain reliable results. To achieve a higher consumption of oxygen and hence a quicker test (30 min), seeding with a known quantity of recirculated sludge has been proposed. The microorganisms are then in high concentration and near the endogenous respiration phase. This second property is important to the application of the proposed respiration technique. As pointed out by Walters et al. [2], the substrate removal rate in an activated sludge system is dependent on the amount of substrate stored within the cells. Thus, a different (higher) removal rate can be obtained when the sludge is properly stabilized (having had time to consume the stored substrate) before contacting it with the influent wastewater. The same phenomenon has been reported by others [3,4] when observing a dampening of the oxygen uptake response because of substrate storage within the cells.

Experimental Conditions

The experimental setup is shown in Figure 1 and consists of an automatic respirometer (Tech-Line Instruments, Arthur Technology, Inc.). It is basically a closed air recirculation system with an air pump, a fine-bubble air diffuser, a 1- to 4-liter aeration column and a CO_2 scrubber [5].

Once filled with the liquor to be analyzed the aeration column is sealed off and the dissolved oxygen uptake rate measurement begins. The pump continuously recirculates air through the aeration column and the scrubber filled with a KOH solution. Fine bubbles rising from the air diffuser ensure adequate mixing but also provide for oxygenation of the liquor. Dissolved oxygen is consumed for the biooxidation of the organic substrate by the microorganisms with an equimolar production of CO_2. Since CO_2 is scrubbed out of the closed air recirculation system, a net decrease in total pressure results from the utilization of oxygen. This change is sensed by a transducer which converts the information to an

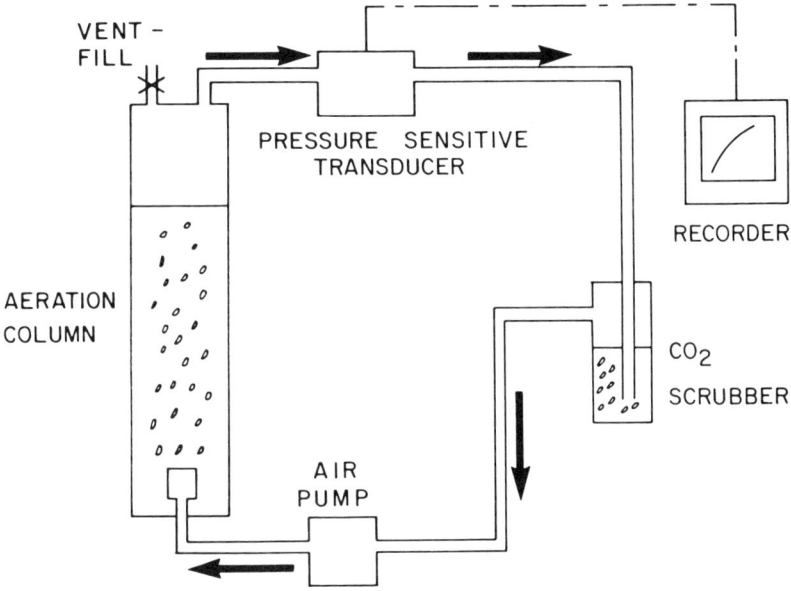

Figure 1. Automatic respirometer.

electrical signal for a recorder. When oxygen falls below a prescribed level, the column is automatically vented for a short period of time to replenish it with fresh air, and the oxygen uptake rate measurement is continued.

Experiments were conducted combining separately domestic sewage and industrial wastewater of prescribed concentrations with known quantities of microorganisms. Activated sludges from the respective treatment plants were used. In each case the seed originated from the sludges being recirculated from the clarification unit to the aeration basin. The samples were collected at a nearby domestic sewage treatment plant and a textile dyeing wastewater treatment plant; both are conventional activated sludge wastewater treatment plants. Some of the characteristics of this industrial liquor are indicated in Table I. They show that the original pH of the liquor was very high, with a low BOD_5/COD (chemical oxygen demand) ratio, both favoring low biodegradability. Also, the initial nitrogen and phosphorus contents were too low to sustain efficient biodegradability. Salts of these elements were added before the test with the pH adjusted to a value of 7.0 Most of the sludges collected on a daily basis at the plants were in a very active stage of respiration far from the expected endogenous stage. Consequently, periods varying from 8 to 15 hr were necessary to stabilize the endogenous level of respiration. During that period the sludges were continously aerated and maintained at room temperature. The sludges were assumed to be in the endogenous state when oxygen uptake reached a plateau.

Due to rapid initial oxygen uptake noted in the early stages of a test, 2 min were allowed between contact of sludges with liquor in the aeration column and the start of measurement. Each test lasted 30 min. Size selection (1–4 liters) of the aeration column was done according to the

Table I. Typical Parameter Values for the Industrial Liquor

Parameter	Value
pH	11.5
BOD_5	458 mg/l
COD	1141 mg/l
Suspended solids	212 mg/l
Total C	220 mg/l
Inorganic C	109 mg/l
Initial N	4.9 mg/l
Added N	18.1 mg/l
Initial P	1.5 mg/l
Added P	3.3 mg/l
Cr^{6+}	3.8 mg/l
Cu^{2+}	0.4 mg/l

VSS content of the sludges. The smaller columns (1–2 liters) were used for sludges having a VSS content above 2500 mg/l. BOD$_5$ and VSS analyses were done according to *Standard Methods* [6].

Experimental Design

A set of daily experiments consisted of a fixed number of 30-min cycles contacting sludges in the endogenous respiration stage, with organic substrate of various strengths obtained by dilution of the original sample collected at the plant. Each set of experiments was run with a freshly collected sample of liquor and sludges.

The oxygen uptake rates associated with the fast initial substrate biosorption (i.e., the total metabolic respiration R_T) and the endogenous respiration rate R_E were measured in each case. Early experiments done with beef extract as a substrate had indicated substrate concentrations for which the biomass (sludges) limited the respiration process [7]. Under such conditions the oxygen uptake rate R_P provided an indirect measurement of active biomass. It has been measured for each set of experiments by contacting the sludges in the endogenous stage with beef extract of concentrations in excess of 1000 mg/l and measuring the oxygen uptake rate over a period of 10 min. This shorter test period has been shown to be adequate for this measurement [7] and was retained here. The quantities of oxygen consumned during a test were denoted respectively $\overline{R}_T$, $\overline{R}_E$ and $\overline{R}_P$, representing the integration of the oxygen uptake rates R_T, R_E and R_P over periods of 30, 30 and 10 min.

In each case the oxygen uptake rate R_S solely due to the substrate assimilation was computed as

$$R_S = R_T - R_E$$

This rate is dependent on the biomass present during the experiment and the specific oxygen uptake rate (SOUR):

$$SOUR = R_S / \text{biomass}$$

This was computed in three different ways, relating biomass to VSS, $\overline{R}_E$ and $\overline{R}_p$.

RESULTS

Table II shows typical values measured for VSS, $\overline{R}_E$ and $\overline{R}_p$ during experiments with sludges collected at intervals of several days. Triplicate measurements were taken in each case and the average reported. The

Table II. Biomass-Related Parameters

Wastewater	Date of Sampling	VSS (mg/l)	$\overline{R}_E$ (mgO$_2$/l)	$\overline{R}_P$ (mgO$_2$/l)
Industrial	17-11-81	3580	4.67	30.78
	20-11-81	6360	12.23	62.97
	21-11-81	5850	7.43	76.39
Domestic	08-10-81	1450	5.11	10.75
	16-10-81	1750	5.85	14.03
	23-10-81	1650	10.93	13.08

results indicate that neither $\overline{R}_E$ and $\overline{R}_P$, both indicative of biomass, is strictly proportional to the VSS content. This agrees with the fact that the VSS content is not necessarily a constant fraction of biomass over time. Also, discrepancies between the trends of $\overline{R}_E$ vs $\overline{R}_P$ are consistent with observations done elsewhere [4,8,9]. Live (hence respiring) but nonviable cells (not active in assimilation and growth) among a population of viable cells may produce such behavior when submitted to widely different experimental conditions (starving for $\overline{R}_E$ vs unlimited food availability for $\overline{R}_P$).

Figure 2 shows the respiration rates R_S for samples of domestic wastewater of various strengths over the time period set for the test. Particularities observed here and for a biodegradable substrate consisting of beef extract [7] or textile dyeing wastewater are listed below:

1. Under conditions where biomass does not limit the assimilation process, both the peak for R_S and its integrated value R_S are proportional to the BOD$_5$ of the liquor.
2. the peak value for R_S always occurs early in the test (4-10 min).
3. As the BOD$_5$ is increased, biomass will eventually limit assimilation, resulting in an asymptotic value for R_S.

When this situation fully develops, $\overline{R}_S$ is no longer proportional to the BOD$_5$ liquor. However, this observation does not preclude the use of this technique for the determination of higher BOD$_5$ liquor. It simply indicates conditions under which dilution is required before testing.

Figure 3 shows the specific respirometric oxygen demand as a function of the specific BOD$_5$ loading using VSS as a biomass-related parameter for domestic wastewater. The specific BOD$_5$ loading used here corresponds to the BOD$_5$ of the liquor per unit biomass. Figures 4 and 5 illustrate the same quantities using, respectively, $\overline{R}_E$ and $\overline{R}_P$ as biomass-related parameters. The data group together results with sludge and substrate samples collected and tested one week apart. This may explain some of the variability noted there. Also, particulate organic material was present in the influent prior to seeding. The test done with the respirom-

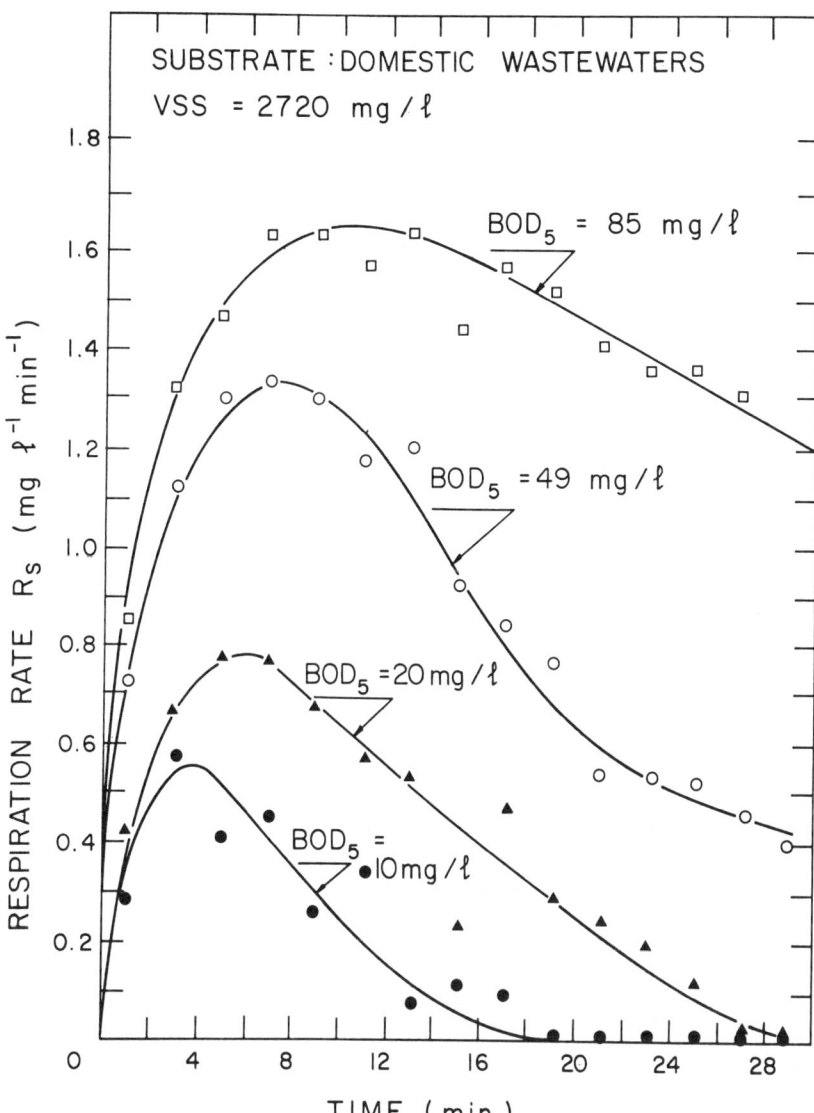

Figure 2. Respiration rate as a function of time and substrate concentration.

eter would then have a better distribution of solids than the BOD_5 test done on a much smaller scale. In each case the data show a trend which is not strictly linear and suggest an asymptotic trajectory for the respirometric oxygen demand ratio with increasing loading. This trend agrees with previous results [7] obtained with highly concentrated beef extract

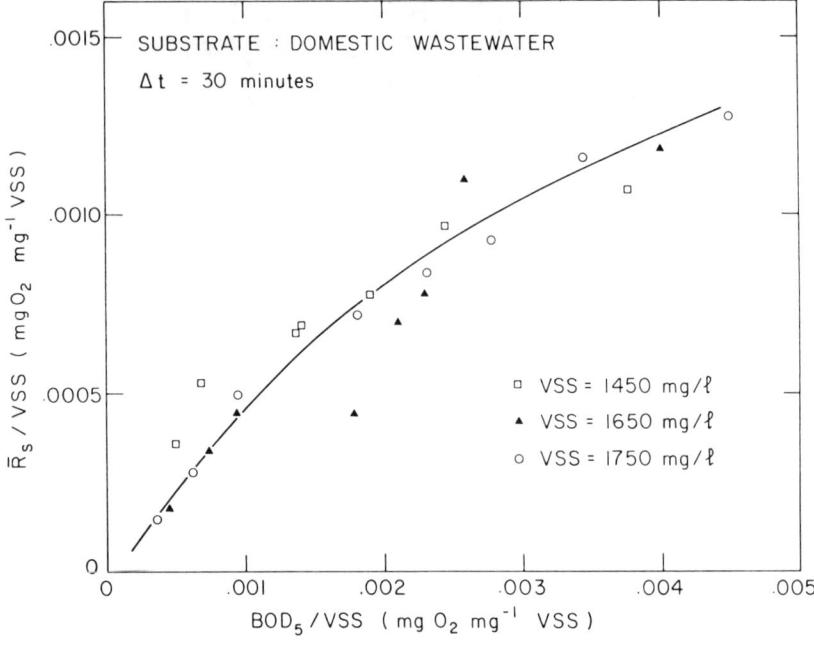

Figure 3. Specific respirometric oxygen demand vs specific BOD_5 loading using VSS as a biomass-related parameter.

liquor. The spread of data in all cases is similar except when using $\overline{R}_P$ as a biomass-related parameter and for which a slight but insignificant improvement is observed over using VSS or $\overline{R}_E$.

Figures 6, 7 and 8 illustrate the same quantities for the industrial wastewater. There is less variability in the data than for domestic sewage regardless of whether VSS, $\overline{R}_E$ or $\overline{R}_P$ is used as a biomass-related parameter. Factors contributing to these results are listed below:

1. The liquor sampled at the plant contained less solids than its domestic counterpart.
2. Te data shown grouped together results obtained with sludge and substrate samples collected and tested daily rather than weekly.
3. Concentrations of the sludges were much higher, thus resulting in more definite measurements.

Again the spread of the data is very similar in each case, with $\overline{R}_P$ again better than $\overline{R}_E$ and VSS as a biomass-related parameter. Although a linear relationship may well approximate the phenomenon over restricted

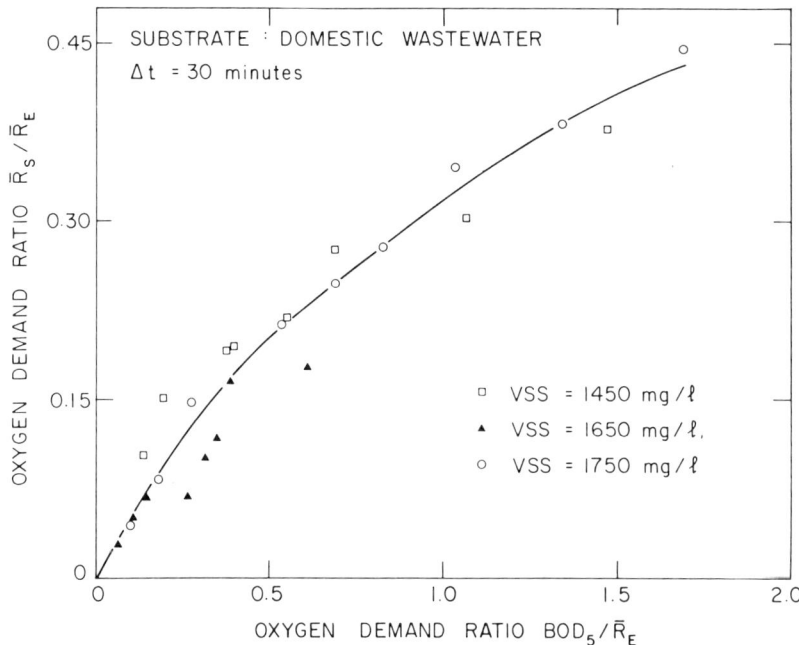

Figure 4. Specific respirometric oxygen demand vs specific BOD_5 loading using $\overline{R}_E$ as a biomass-related parameter.

ranges of BOD_5 loadings, the overall trend is nonlinear, confirming more definitively the asymptotic behavior observed before.

When considering the whole range of substrate concentrations the relationship obtained for each of the liquors studies is

$$R_S^* = \frac{C_3 \times BOD_5^*}{C_4 + BOD_5^*}$$

a formulation of the Michaelis-Menten form, where

$$R_S^* \; \overline{R}_S/(\text{biomass-related parameter})$$
$$BOD_5^* = BOD_5/(\text{biomass-related parameter})$$

However, a linear relationship is essentially observed for the range of substrate concentrations when biomass is not limiting the assimilation process, giving

$$R_S^* = C_1 \times BOD_5^* + C_2$$

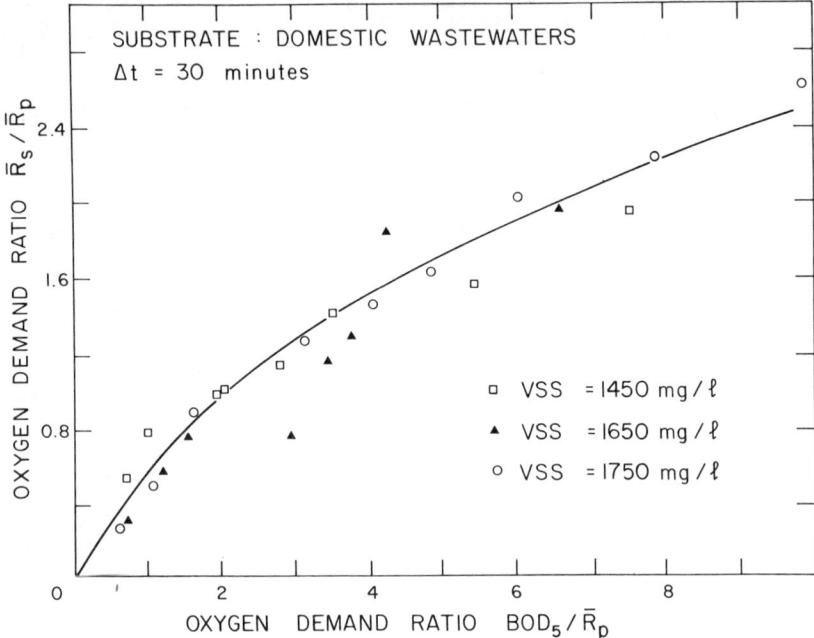

Figure 5. Specific respirometric oxygen demand vs specific BOD_5 loading using $\overline{R}_P$ as a biomass-related parameter.

A Continuous Experiment

To test the validity of using an explicit Relationship between R_S^* and BOD_5^* for predicting the BOD_5 of a liquor, the following experiment was designed. Wastewater entering the treatment plant was periodically sampled over a period of 24 hr. Each sample was properly seeded with an acclimated sludge already in the endogenous phase and the respirometric test was run for 30 min. Therefore 48 samples were analyzed in the time period. To further check the predictions, the BOD_5 (triplicata) was done on each of these samples. The $\overline{R}_P$ parameter has been used in relating it to biomass. The data are shown in Figure 9 with the solid line representing the best fit using the Michaelis-Menten formulation.

Table III indicates the corresponding set of best-fit coefficients for the other biomass-related parameters, as well as for the linear formulation. Computation of the BOD_5 for each of the 48 samples was done using both formulations for the three biomass-related parameters VSS, $\overline{R}_E$ and $\overline{R}_P$. The mean relative error and standard deviation of the relative error for

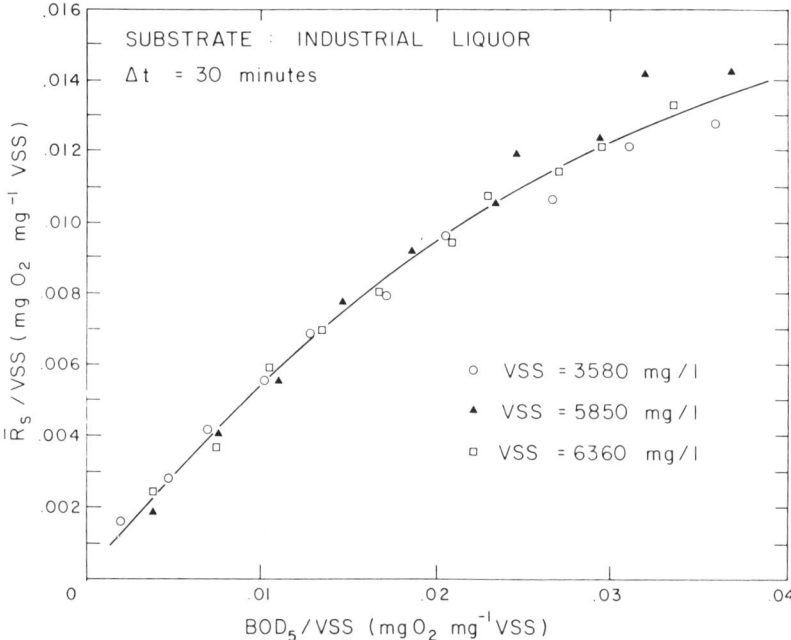

Figure 6. Specific respirometric oxygen demand vs specific BOD_5 loading using VSS as a biomass-related parameter.

the set of 48 samples are reported in Table IV. Note in Figure 9 that a linear formulation may give a satisfactory estimate with a mean relative error averaging 10.2% and a corresponding average standard deviation of 7.2%. For the Michaelis-Menten formulation a slightly better estimate may be obtained with a mean relative error averaging 9.6% and a corresponding average standard deviation of 6.8%.

Except for the case when $\overline{R}_E$ was used to characterize biomass, the Michaelis-Menten formulation was superior to the linear formulation. Also, using $\overline{R}_P$ as the biomass-related parameter consistently produced better results for both formulations, showing a reduction of roughly 10% in the error as compared to using VSS or $\overline{R}_F$.

Finally, Figure 10 illustrates the comparison of the BOD_5 obtained by direct measurement (5 days later) and the predicted BOD_5 values using the specific oxygen uptake rate with the Michaelis-Menten formulation and $\overline{R}_P$ as the biomass-related parameter. The BOD_5 trend with time is well predicted and represents an adequate estimate of the organic loading entering the wastewater treatment plant.

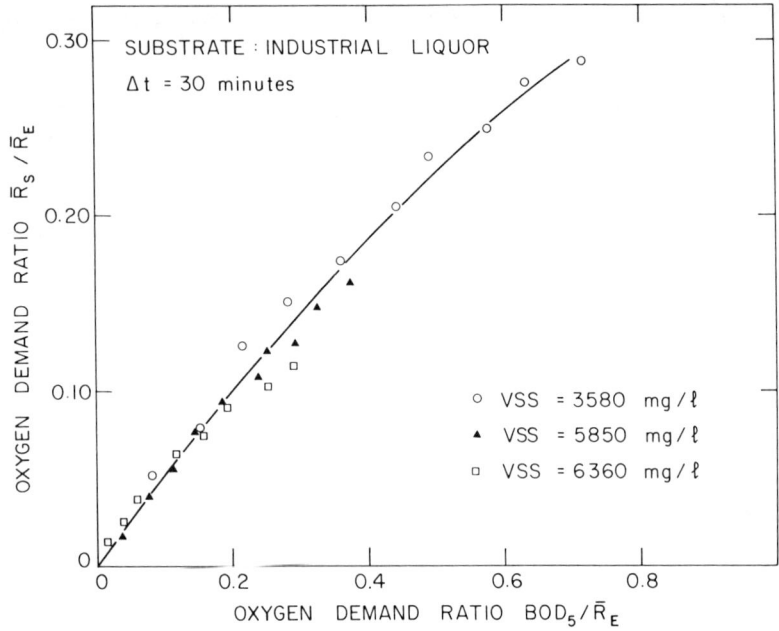

Figure 7. Specific respirometric oxygen demand vs specific BOD_5 loading using $\overline{R}_E$ as a biomass-related parameter.

IMPLICATIONS FOR CONTROL

An on-line respirometer can provide the information discussed above after each cyclic measurement of $\overline{R}_S$. Typically the predicted value of BOD_5 entering the plant can be obtained after a time delay of 30 min or less. This represents a quasi-instantaneous measurement when considering the commonly large time constant of these plants.

A routine BOD_5 determination on samples fed to the respirometer can then permit the continuous updating of a calibration curve based on the most recent data taken. Such a curve, with the output originating from an on-line respirometer, will permit the automatic monitoring of the biodegradable organics entering an activated sludge wastewater treatment plant. This information can then constitute the input to a control system for the plant.

A point of practical interest to be noted here is that on-line BOD_5 estimation using the concept of $\overline{R}_P$ (or $\overline{R}_E$ as a biomass-related parameter

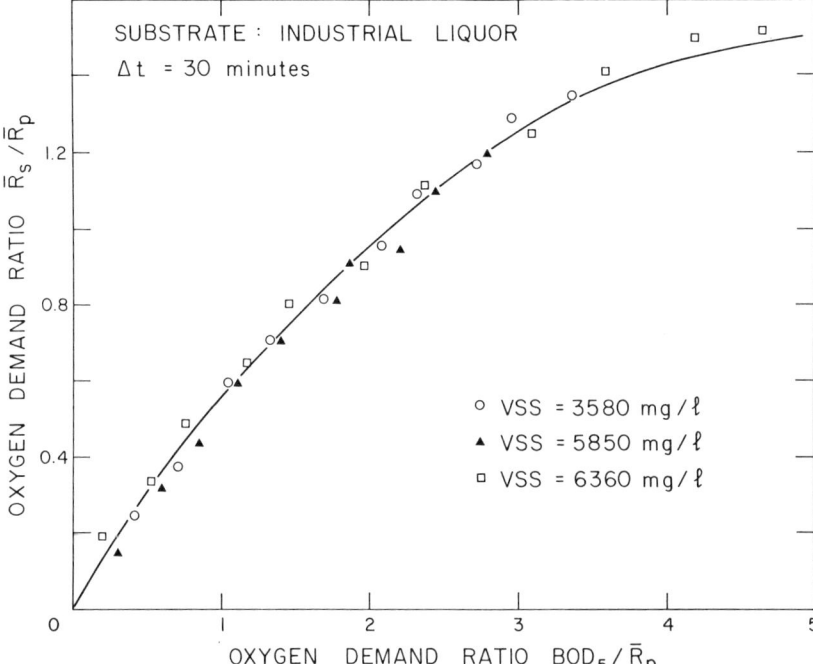

Figure 8. Specific respirometric oxygen demand vs specific BOD_5 loading using $\overline{R}_P$ as a biomass-related parameter.

avoids the problems of on-line determination of VSS (usually SS) for the computation of the specific oxygen uptake rate. Also, separate measurements of $\overline{R}_P$ (or $\overline{R}_E$) need be done only once or twice a day to provide a reliable estimate.

SUMMARY

The general objective of this study was to discuss conditions under which reliable short-term estimates can be obtained for the BOD_5 of biodegradable wastewater using automatic respirometers. More specifically, $\overline{R}_P$ (the oxygen uptake rate when biomass is limiting the substrate assimilation process) has been used and compared with other biomass-related parameters such as VSS and $\overline{R}_E$. Its use in relating the specific oxygen uptake rate to the BOD_5 of domestic and industrial wastewater has produced results consistently showing less variability.

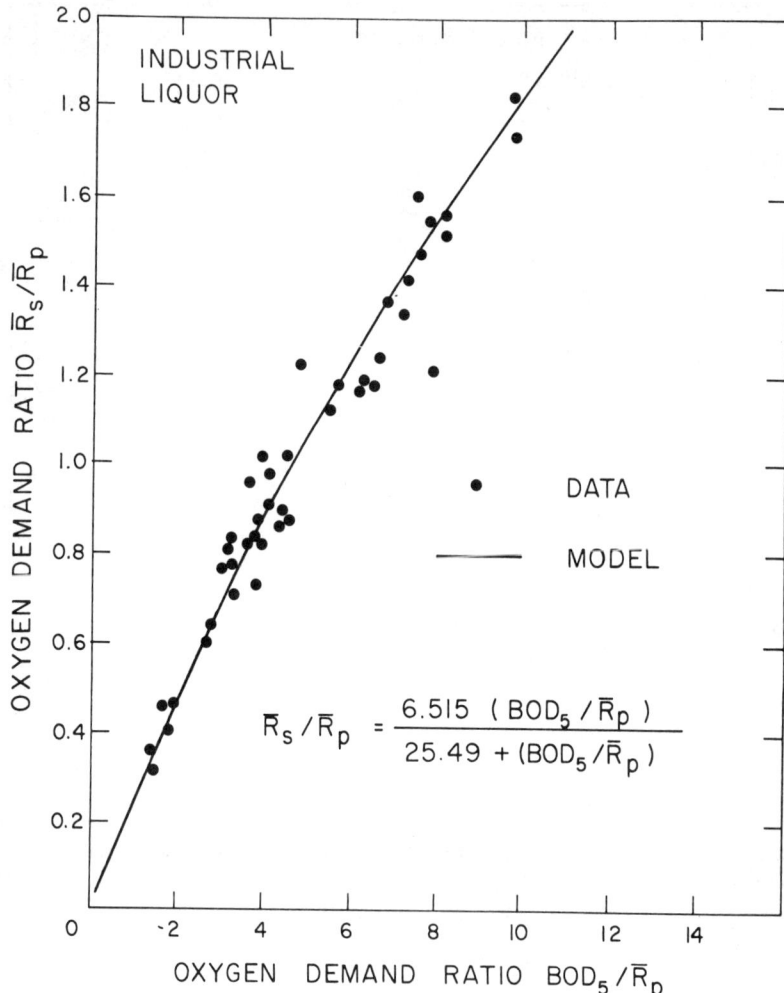

Figure 9. Results for the continuous experiment.

A second outcome of the study was to indicate that a linear relationship between specific oxygen uptake rates R_S^* and specific BOD_5 loadings is observed when biomass is not limiting the substrate assimilation process. Results have also shown that a Michaelis-Menten formulation fits the data well and produces reliable BOD_5 estimates even when biomass influences the substrate assimilation process.

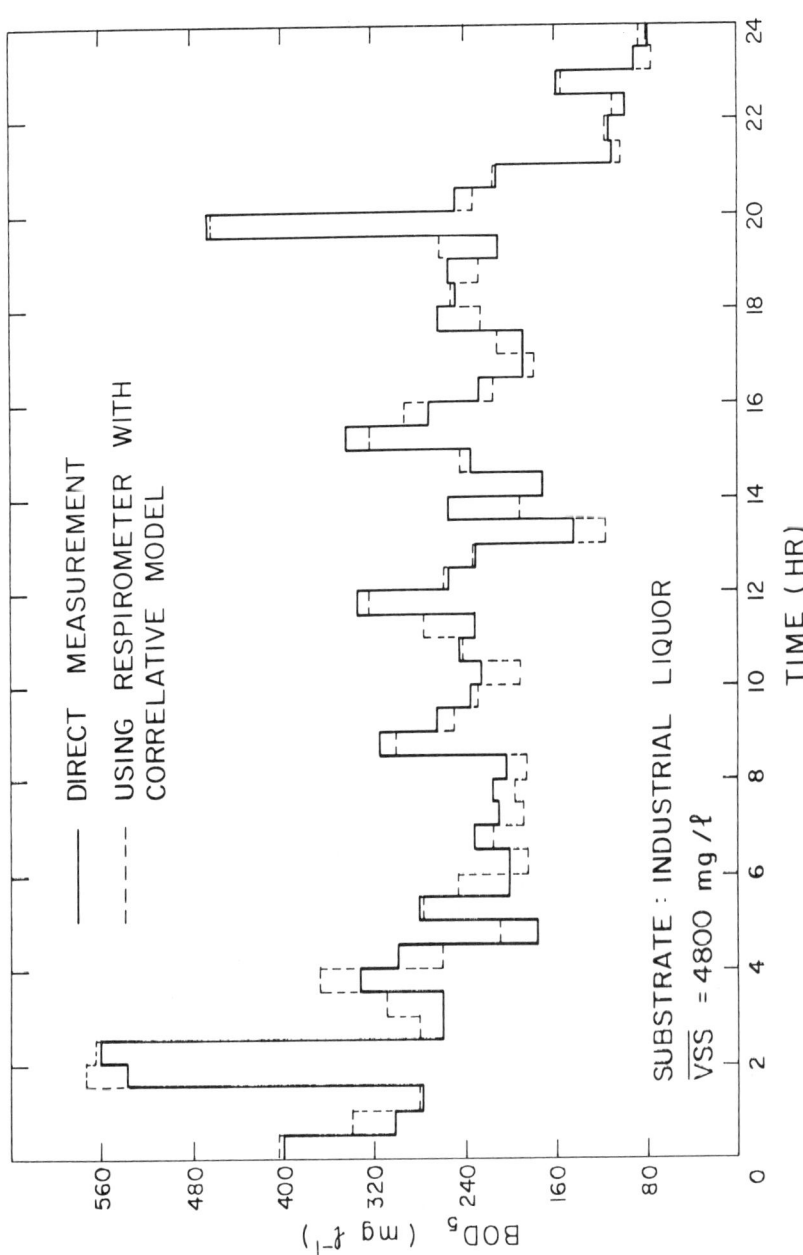

Figure 10. Comparison between predicted and measured BOD_5 at the plant.

Table III. Coefficients for the Linear and Michaelis-Menten Formulations

R_S^*	BOD_5^*	C_1	C_2	C_3	C_4
$\overline{R}_S/VSS$	BOD_5/VSS	0.19865	0.00053	0.06038	0.2357
$\overline{R}_S/\overline{R}_E$	$BOD_5/\overline{R}_E$	0.19002	0.80971	40.5820	148.89
$\overline{R}_S/\overline{R}_P$	$BOD_5/\overline{R}_P$	0.19187	0.08218	6.51500	25.490

Table IV. Relative Error and Standard Deviation Made Using the Predictive Models

Biomass Characterization	Linear Form		Michaelis-Menten Form	
	$\overline{X}$ (%)	S (%)	$\overline{X}$ (%)	S (%)
VSS	10.69	7.29	9.35	6.57
$\overline{R}_E$	10.15	6.91	11.04	7.17
$\overline{R}_P$	9.69	7.51	8.56	6.81

Finally, the results are of significance in view of process monitoring and control. They provide alternative and more accurate ways to predict organic loadings to activated sludge wastewater treatment plants.

For further information on this topic, see the Questions and Answers section on page 217.

References

1. Arthur, R.M., and W.N. Hursta, "Short Term BOD Using Automatic Respirometer," in *Proceedings of the 23rd Purdue Industrial Waste Conference* (West Lafayette, IN: Purdue University, 1968), pp. 242-250.
2. Walters, C.F., R.S. Engelbrecht and R.E. Speece, "Microbial Substrate Storage in Activated Sludge," *J. San. Eng. Div. (ASCE)* 94(SA2):257-269 (1968).
3. Cufft, R.C., and J.F. Andrews, "Predicting the Dynamics of Oxygen Utilization in Activated Sludge Process," *J. Water Poll. Control Fed.* 53(7):1219-1232 (1981).
4. Duggan, J.B., and J.L. Cleasby, "Effect of Variable Loading on Oxygen Uptake," *J. Water Poll. Control Fed.* 48(3):540-550 (1976).
5. Tech Line Instruments, Arthur Technology. *Laboratory Wastewater Respirometer Application Book*, Fond du Lac, WI (1976).
6. *Standard Methods for Examination of Water and Wastewater*, 15th ed. (New York: American Public Health Association, American Water Works Association and Water Pollution Control Federation, 1981).

7. Ilhan, F., and N. Thérien, "Influence of Biomass-Related Parameters for BOD_5 Determination Using Automatic Respirometers," *Water Res.* (Submitted for publication, 1982).
8. Mikesell, R.D. "Aerobic Bacterial Metabolic Process," *J. Environ. Eng. Div.* 107(EE6):1261–1276 (1981).
9. Grady, C.P.L., and R.E. Roper, "A Model for the Bio-Oxidation Process which Incorporates the Viability Concept," *Water Res.* 8:471–483 (1974).

CHAPTER 9

DESIGN OF ACTIVATED SLUDGE FACILITIES FOR OPERABILITY AND MAINTAINABILITY

Vernon T. Stack
  Smokey Stack, Inc.
  Cedars, Pennsylvania

Edward L. Gillette, Jr.
  Betz, Converse, Murdoch, Inc.
  Plymouth Meeting, Pennsylvania

INTRODUCTION

The concepts presented in this chapter are the collective opinions and points of view of (1) an engineer who is experienced as a department head responsible for concept design, startup, and operations projects for wastewater treatment facilities, and (2) an engineer who has many years of experience in the operations and maintenance (O&M) of wastewater treatment facilities.

The perspective taken is applicable to activated sludge facilities for treatment of domestic wastewaters. However, most of the points discussed are relevant to industrial wastewater treatment facilities. A major difference is that the industrial owner is more deeply involved in the concept and detailed design of a facility.

Our objective is to identify some of the concept and detailed design decisions which can create O&M problems. A more informed view of such problems on the part of the engineer, owner, and regulatory agency is definitely needed.

POINTS OF VIEW

The physical makeup of a wastewater treatment facility is shaped by the opinions of a "committee of architects" that includes the owner, engineer, operating personnel and regulatory agencies. Distillation of the points of view of the "committee" into the ultimate facility is a difficult task which falls mostly to the engineer. Since the decisions which shape the design are obviously important to operability and maintainability, a brief examination of the members of the "committee" is appropriate.

The Owner

Municipal owners have characteristics which to a degree are related to the size of the municipality, although the "personality" of the owner may be related to the individual(s) who have specific interests and campaign for those interests. Larger municipal owners generally have a staff of career personnel who give the owner a stable orientation to engineering projects. The general assumptions of the owner of large municipally-owned treatment works are listed below:

1. The staff will be involved in the conceptual and detailed design of the facility.
2. The staff will be sensitive to capital and operating costs.
3. The staff has a marginal interest in innovative technology because the administration of a municipality is not in a position to take risks.
4. The design life of the facility must be long (20 yr or longer) because the funding is long-term.
5. The performance of the O&M staff will be marginal to acceptable.
6. The owner will consider sophisticated process instrumentation and controls, but has limited staff capability to maintain such systems.

At the opposite end of the spectrum from the large municipalities are the smaller municipalities which do not have professional staffs. The governing body, such as a mayor or manager and council, are faced with a project in which their perspective is predominantly political. The governing body is dependent on the engineer for technical guidance. The elapsed time between the design, construction and startup of a wastewater treatment project may extend over one or more changes in the administration of the governing body.

There are exceptions, but generally, the small municipality has only a limited ability to develop and keep within the governing body a firm knowledge of, and respect for, the O&M requirements of a wastewater treatment facility. The small municipality also may find it difficult to develop and retain a quality O&M staff. Thus, it is generally true that smaller municipalities should not be saddled with wastewater treatment processes that require careful operation and maintenance.

In characterizing the small municipal owner, the following should be noted:

1. Although limited in technical knowledge, the small municipal owner is sensitive to capital and operating costs and may force design limitations that make a facility more difficult to maintain or operate.
2. The small municipal owner probably will have a limited input to the conceptual and detailed design of the facility, and will depend on the consulting engineer's recommendations.
3. The small municipal owner probably will hire fewer than the recommended number of operating personnel and will have difficulty in attracting and keeping quality personnel because of the low salary scale mandated by the small size of the municipality.

The Engineer

The engineer, as a collective organization, has many objectives that require attention. The predominant one is the selection and implementation of a practical facility design for the particular requirements of the client with minimum capital and operating costs. The individual personnel who do the conceptual and detailed design most likely do not have significant experience in the actual opertation of wastewater treatment facilities. Many years of design experience may do little to improve a design engineer's understanding of O&M problems if the tasks of startup and operations assistance are done by another group within the firm, such as the operations department.

Within the same engineering organization, both the conceptual and detailed designs should be reviewed by operations personnel. However, even with a good operations review, it is difficult to have the O&M point of view represented adequately in the final design. It is difficult to assign the correct value of the operations effort generated by a particular design decision. The cost of the equipment or structures is easily understood, but it is more difficult to quantify the difficulty or inconvenience, associated with a particular O&M effort, which increases the hours of operator effort or attention. As a result, the engineering decisions may create O&M problems that are not recognized until startup and operation of the facility. Engineering organizations need to make a greater effort during the design stage to understand the real scope of operations and maintenance difficulties.

Operations Personnel

Within the owner's staff there may be personnel, such as a plant superintendent, who can make a significant contribution to an operability and maintainability review of the conceptual and detailed design. If the project is for an industrial client, the input of the owner's personnel will be provided with clear emphasis. If the project is for a municipal client, the input of the client's operations personnel may not be given the same importance, or may be displaced more easily by the obvious capital costs. Again, the engineer should attempt to give suitable weight to capital O&M cost factors.

Regulatory Agencies

The review of construction drawings and documents by regulatory agencies seldom considers the O&M aspect of a project in any significant depth. The normal operational considerations are for power, chemicals, and personnel to a limited extent, with very little attention given to the possible difficulties in operations and maintenance. An exception may exist when a project is given a review for "value engineering." Generally, the correct attention to the operability and maintainability question must be addressed by the engineer and the owner together.

The position taken by the regulatory agency during review of conceptual design for a facility is important because it also may lead to O&M difficulties. The engineer develops a design basis and provides a conceptual design that will meet the effluent limitations when the entire facility is operational. A basic requirement by the regulatory agency is that there must be parallel units in the facility. For small facilities, the resulting concept design is probably two parallel units for each unit operation that the regulatory agency finds acceptable. Stated below are two obvious questions that are pertinent and are not considered seriously at the time of review.

1. If a unit operation is out of service, how will the process perform?
2. What will be the quality of effluent produced?

Generally, the view of the regulatory agency during concept review is that the process is conceptually "operational" with one-half of a unit operation in service. Tacit approval is given for reduced effluent quality.

Identifying the interplay between the engineer and regulatory agency at the time of the conceptual design is not an indictment of either party. They both are correct in giving facility costs prime importance. Standby tankage is not justified, and the use of multiple units also increases costs. The two-parallel-unit approach appears to meet all objectives. Nevertheless, from the operational point of view, it is not clear that the objectives have been stated and examined carefully enough.

Once the facility is in operation, however, the posture of the regulatory agency changes. Enforcement personnel are not inclined to accept an effluent quality less than the National Pollutant Discharge Elimination Standards (NPDES) limit merely because part of an operating unit is out of service. Thus, the owner and operations personnel are placed under stress by well-intentioned, but questionable, cost-related decisions on the part of the engineer, and the tacit but unofficial approval given by the regulatory agency at the time of the initial design review.

FACILITY DESIGN CONSIDERATIONS

Process Flexibility

Process flexibility means that, within the tanks and piping provided, there are opportunities to make alterations that benefit operations by increasing process stability, reducing power or the cost of chemicals and saving operating effort. Following is an example:

A. Within the facilities plan for a domestic facility, estimates are made of projected increases in utilization of a facility during its operating life. From this information, an estimate of the size of a facility required to treat the wastewater flow efficiently during periods of the plant's operating life can be made. If the quantity of wastewater will only be one-third of the plant's ultimate capacity for several years, the facility should have the ability to handle only one-third of the flow and to do so economically. The parallel unit operations possibly should be designed for one-third or one-fourth capacity.

If cost considerations lead to the selection of parallel units at one-half of the ultimate capacity, this can result in operational headaches. Oversized clarifiers provide holding times that are too long and may cause release of soluble organics to the effluent. Overoxidation of biomass may cause pin floc loss to the effluent. Generally speaking, sludge management may be difficult if facilities for returned sludge and sludge wasting are not suitably sized.

In an industrial project, consideration may be given to phased construction of facilities in order to stage the capital investment. In municipal projects, staged construction has not been practiced because EPA construction grants have been based on the total project. With the pending changes in availability of grants, however, staged funding may become feasible.

In any event, the conceptual design of activated sludge facilities should consider carefully the use of an adequate number of parallel units in the process so that the plant can be operated effectively with minimum operating costs. Cost considerations to be examined include power costs for mixing and for autooxidation of sludge, cost of chlorine and other chemicals and reduction of overtime for operating personnel.

B. The activated sludge process should be designed for several possible modes of operation, including plug flow, complete-mix, step feed and contact stabilization. The engineer does not have a clear perspective on the characteristics of the wastewater to be treated. Industries may leave or locate on the municipal system, and the treatability of the wastewater may change as well, when one or more of the following occurs:

1. The fraction of nonsoluble BOD may increase and a contact stabilization may be the best mode,
2. A larger fraction of readily biodegradable organic material may require a complete mix or step feed mode,
3. A need for greater autooxidation of sludge may call for a plug flow mode of operation.

Pretreatment

The basic objective in pretreatment is to adjust the character of the wastewater so that there will be less process or operations and maintenance problems within the subsequent unit operations. Pretreatment operations (not including chemical feeds) may include the following:

- equalization
- grit removal
- rag removal

Equalization

Industrial facilities, rather than municipal treatment works, are more likely to need equalization. Theoretically speaking, in municipal projects, the changes in industrial wastewater caused by pH or slug discharges are handled more cost-effectively at the source, rather than in combination with a large flow of domestic wastewater. Though effective, internal equalization of the activated sludge process cannot be expected to handle abnormal situations.

Pretreatment equalization to level diurnal wastewater loads and any accidental slug loads from industry offers significant advantages in process operations. Process stability is improved, power requirements are reduced, and, conceivably, the conceptual design can allow for smaller units. However, at the time of design, equalization is difficult to justify. Daily changes in effluent quality are not a part of the effluent criteria. An equalization unit would have to have a detention time of approximately 24 hr and the effluent would have to be mixed and aerated; this is an expensive unit.

Later, when the facility is in operation, there will be events such as slug loads or inadequate nutrients during part of the day which could have been corrected in whole or in part by equalization; hence, the wisdom of not providing equalization then will be questioned. This type of problem is generally the result of industrial discharges. In lieu of equalization, a very effective industrial pretreatment program must be implemented.

Grit Removal

Grit removal, as normally practiced utilizing aerated channels, is effective. Obviously, the effective removal of grit is important in reducing the wear on pumps and piping in the sludge management system. In addition to causing wear on equipment, grit has a higher density than sludge and thus is more difficult to suspend. Grit can fill sludge tankage and require a special maintenance effort. In systems without primary clarification, aeration basins will accumulate grit.

Taking one grit removal unit out-of-service should have no measurable impact on the plant's grit removal efficiency. However, meeting this objective probably requires the installation of at least three parallel units. Any two units should be able to handle the daily hydraulic peak.

Screening

Rags in the sludge system can create severe operations and maintenance problems. Comminution of raw sewage can reduce rags to strings, but strings can reweave to form ropes which may clog pipes and pumps. Successful removal of rags requires an effective screening system. Front-raked bar screens are not effective because the raking operation may push rags through the bars. Back-raked bar screens are more effective in rag capture, but the bars should be spaced about one-half inch apart. Sloped static screens also are effective in rag removal and should be considered wherever raw wastewater is pumped into the facility. An out of service screening unit should not reduce the efficiency of the plant's rag removal. Provided there are at least three parallel units, the remaining units should be able to handle the daily hydraulic peak.

Primary Clarification

With effective grit and rag removal in place, the process protection aspects of primary clarification are reduced. If the efficiency of primary solids removal is reduced or eliminated for brief periods, the activated sludge process will not be affected significantly, provided the plant has adequate capacity to handle the increased organic load. On the other hand, the approximately 30% BOD removal achieved by primary clarification is very economical when compared to costs incurred for equivalent removal in the aeration basin. If the design allows for transfer of more BOD to the aeration basin when a primary clarifier is out-of-service, the design also must provide a larger aeration basin, more oxygen transfer capacity, and possibly more secondary clarifier capacity to accommodate the higher BOD load. Thus, the penalty is significant when primary clarification is not provided. A truly economical design should optimize the efficiency of primary clarification. The minimum number of clarifiers probably should be four, with an even larger number preferred.

Primary clarification becomes more important for process protection when the wastewaters contain an abnormally high quantitiy of mineral oil and grease. Clarification capacity should be maintained for release and removal of scum. The capacity with one clarifier out of service should provide an overflow rate of not more than 1000 gpd/ft^2 at the maximum monthly flow. Since the maximum monthly rate in a municipal system is normally about 20% greater than the average monthly rate, the design objective can be achieved with four clarifiers designed for a monthly average overflow rate of 600 gpd/ft^2.

Aeration Basin Capacity

Aeration basin volume required to provide a given effluent quality is related to the design of the secondary clarifiers and the underflow area available. Figures 1 and 2 present an approximate model of aeration volume and clarification area for activated sludge treatment of domestic sewage. Average kinetic and sludge flux rates were utilized. The design modeled is 30 mg/l of effluent BOD$_5$ under winter operation, and, simplistically, any design condition on the Figure 1 and 2 plots theoretically provides the desired effluent quality. Other considerations, such as clarifier overflow rates and sludge residence times in the clarifier, would have to be applied.

The present intention is to illustrate the impact of reduced aeration volume on required sludge recycle capacity. Consider an aeration basin detention time of 7.2 hr (V/Q = 0.3) and clarifiers designed to provide a

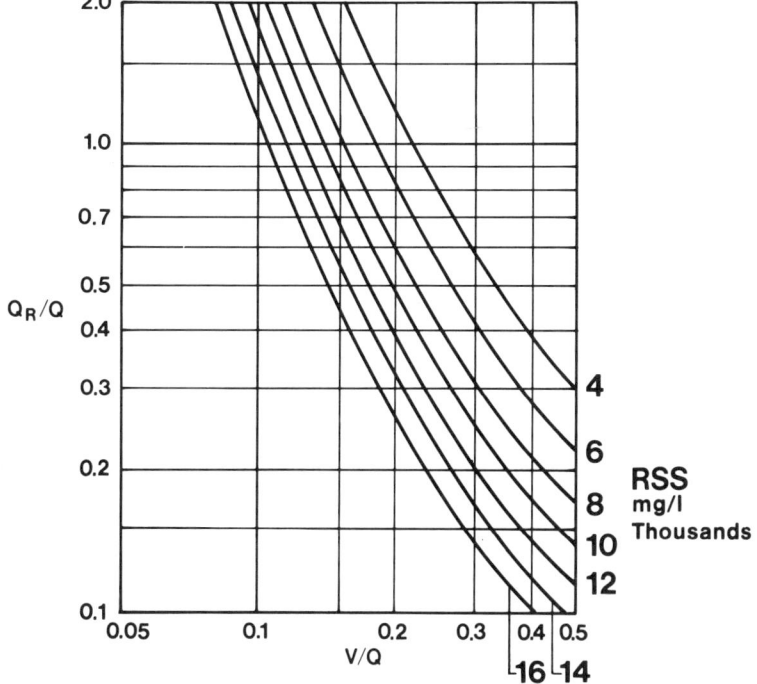

Figure 1. Basin volume and recycle biomass.

returned sludge concentration of 12,000 mg/l. From Figure 2, the underflow area (UA/QP) would be 1900 ft²/MGD.

If the aeration volume is reduced to 50% (V/Q - 0.15), the returned sludge rate (to produce an effluent of 30 mg/l at UA/Q = 1900) results in a recycled sludge concentration, RSS, of approximately 4000 mg/l. From Figure 1, at V/Q = 0.15 and RSS = 4000 mg/l, the required sludge recycle ratio, Q_r/Q, is about 2:1. If the exercise is repeated for other reductions in aeration volume at the selected design, the estimates of recycle ratio available aeration volume shown in Figure 3 are generated.

To give the process the flexibility to meet the effluent criteria when aeration volume is reduced by 50%, a sludge return capacity of 200% would be required. During normal operating conditions, the sludge recycle capacity use would be about 20%. Design of the sludge recycle capacity for the full range presents problems with line velocities and pump capacities. The suggested design approach is to limit the potential

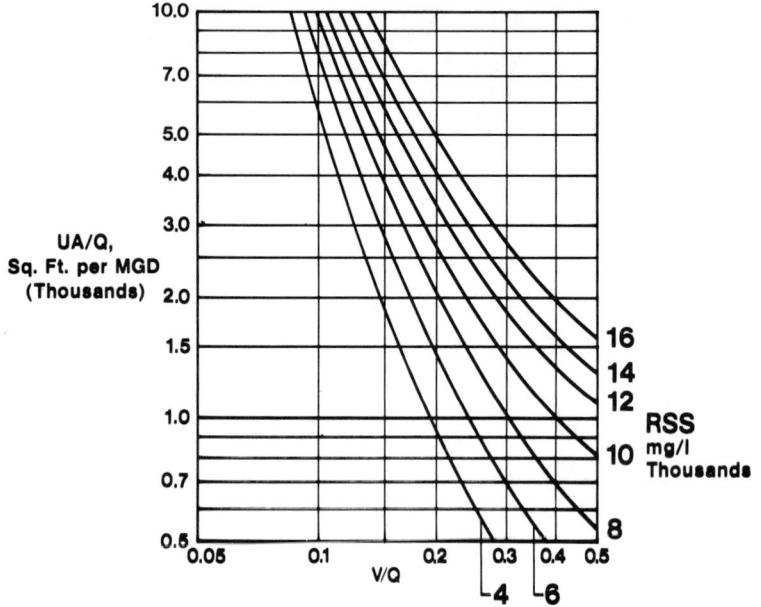

Figure 2. Basin volume and clarifier area.

for loss of aeration volume by providing at least three aeration basins in parallel and a sludge recycle capacity of 100%.

As pointed out previously, regulatory agencies, at the time of concept design review, generally do not require that the process produce the same quality of effluent when part of a unit operation is out of service. Where the compromise in the concept design provides two parallel aeration basins with the total volume sized for the process objective, the design basis has a built-in operational problem, in addition to problems with effluent quality, when one aeration basin is taken out of service. With limitations to the amount of sludge that can be returned, the resulting higher organic loading per unit weight of biomass may result in a predominant growth of filamentous organisms. Prompt action must be taken to correct the growth, or the resulting decrease in sludge flux rate will further compound this problem. Also, if at the time of startup, the facility is treating less than one-half of the design flow, the two-train concept may provide even less flexibility, as discussed earlier.

When the process flexibility problems and posture of regulatory agency enforcement personnel are considered, the two-parallel-unit approach is a poor choice. The minimum number of parallel aeration basins probably

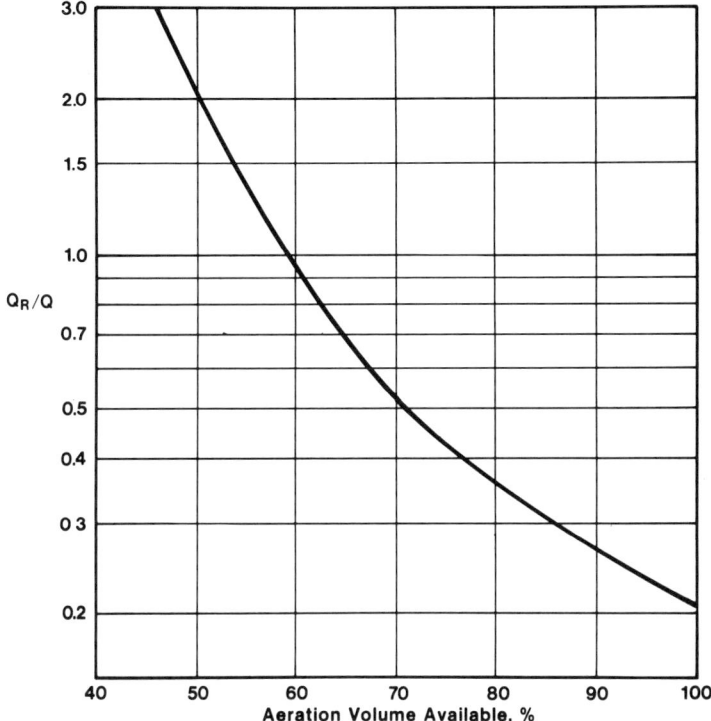

Figure 3. Required recycle capacity.

should be three, or possibly four, if the startup load to the process is 50% or less of the design capacity.

Secondary Clarification

Secondary clarification has the dual functions of biomass capture to achieve a suitable effluent quality and biomass thickening for return to the aeration basin. Based on the process kinetics, either overflow rate or underflow rate will control the sizing of the clarifiers. A factor that must be considered is the amount of time the sludge is in the clarifier. With long residence time, there is a potential for cell lysing and release of organic material to the effluent of the clarifier. Based on the total clarifier volume and the quantity of sludge returned, the theoretical residence time should be limited to approximately 4 hr.

A second factor in clarifier design and operation is assurance that the bottom of the clarifier is cleaned by the sludge removal mechanism. Obviously, sludge not removed will become septic and eventually rise to the surface of the clarifier by action of the generated gases. When bottom suction devices are used, a rubber squeegee in contact with the bottom must follow immediately after the suction orifices. Without the squeegee, the bottom pickup will clean in the vicinity of the orifice, but a mound of sludge several inches high can remain on the bottom between the orifice locations, become septic, and create final effluent problems.

Considering the interrelationship of the secondary clarifier underflow area and the returned sludge necessary for the activated sludge process, two clarifiers designed at a total overflow rate of 600 gpd/ft^2 cannot provide a flexible process. If one clarifier is out of service, the effluent quality must decrease because suspended solids capture is less efficient. Also, the process probably will suffer because the one clarifier probably will be underflow limited, and biomass return to the aeration basins will be limited. The number of secondary clarifiers should be adequate to minimize the impact when one is out of service. If the average overflow rate is conservative—for example, 600 gpd/ft^2—a minimum number of four secondary clarifiers should be considered. If the design is less conservative, particularly for larger facilities, a larger number of clarifiers should be considered.

Return Sludge Capacity

As demonstrated, the aeration volume, secondary clarifier area and return sludge capacity are all interrelated. Within this relationship, settling characteristics and the resulting sludge flux rate are variables. At those times when the settling rate of the activated sludge decreases, it is vital that return sludge pumping capacity be available so that a higher rate of return sludge can be implemented to keep the quantity of sludge in the aeration as optimal as possible. Based on forward flow, the return sludge rate should be at least 100%; an even higher rate possibly should be considered. When needed, higher sludge return rates result in higher sludge flux rates (at lower RSS concentrations) from the secondary clarifier.

Aeration Capacity

There is no substitute for having adequate oxygen transfer capacity where it is needed, and it may be difficult to provide the quantity and range of transfer when the number of aeration basins is limited. For example, for the more efficient oxygen transfer devices, such as fine-bubble diffusers, oxygen transfer efficiency decreases with the power density. If one of two parallel aeration basins is taken out of service, placing all of the available air into the remaining basin will not provide the same transfer capacity as that which existed in the two basins; transfer capacity will be approximately 15 to 20% less.

Providing an adequate number of diffusers in one basin to handle approximately twice the air flow also carries penalties. In addition to having possibly 50% more diffusers per tank, the minimum airflow required per diffuser limits the low airflow when both basins are in service. Thus, there is a penalty for capital investment as well as operating power. When examined carefully, the operational factors related to having only two aeration basins in parallel favor a larger number of parallel units.

Chlorination

As normally applied, chlorination system design is satisfactory. The lines from the chlorine cylinders to the chlorinator should present no problem, if moisture does not enter the system. In actual operation, however, moisture does enter the system and the chlorine lines may become clogged with ferric chloride. To prevent this from occurring, all chlorine lines should be kept as short as possible. Periods of poor secondary clarifier performance and/or improved settling with chlorination may cause solids in the secondary clarifier effluent to accumulate in the chlorination basin. The basin should be provided with bottom scrapers for convenient removal of sludge.

Processing of Side Streams

Processing of sludges can result in the return of significant volumes of high-strength wastewater to the activated sludge process. The magnitude of the side stream recycle load depends on the process used for sludge stabilization.

Aerobic stabilization results in side streams from thickening and filtration operations. The BOD of these streams is relatively low (generally a few hundred milligrams per liter). Most of the ammonia present has been oxidized to nitrates. Return of the nitrates to the head of the plant represents a chemical source of oxygen that can be beneficial. The aerobic sludge digestion process has a minimal impact on the activated sludge process.

Anaerobic stabilization of sludges can result in the return of high-strength side streams to the activated sludge process. Two-stage high-rate anaerobic digesters generally have been designed on the concept of removing a supernatant from the second stage. In actual practice, the supernatant generally is very high in solids (1 to 2%); the impact on the process is significant regardless of where the supernatant is returned in the activated sludge system. There are two approaches that can resolve this problem:

1. The side stream from supernating the digesters or filtration of the sludge can be stabilized aerobically before return.
2. A special effort can be made to thicken the sludges before digestion so that digester supernating is not required.

When compared to the digester supernatant side stream, the side streams from sludge thickening contain significantly less BOD and solids.

For the anaerobic sludge digestion process, adequate thickening of the sludge streams (to a minimum of 6% solids) is the better design approach). In addition to returning less BOD and solids to the activated sludge process, thickening can reduce digester size requirements and lower the amount of energy required for digester heating. The filtrate from dewatering of anaerobically digested sludge should be equalized and aerated before return to the activated sludge process.

Very difficult side streams are those from heat treatment processes that solubilize protein in the sludge without providing oxidation. Supernatants or filtrates from this kind of process can have a BOD of about 10,000 mg/l and an extremely obnoxious odor. The side streams may represent as much as 25% of the total load to the activated sludge process and should be treated before return to the process. A treatment approach is to provide a dedicated oxygen activated sludge facility with a covered reactor to reduce the load and to control the odor. Because of the odor and high BOD problems involved, heat treatment of sludges is a questionable alternative for sludge conditioning or stabilization.

CHEMICAL FEEDS

Chemicals are tools which may be applied to improve the performance of unit operations.

Pretreatment and/or Primary Clarification

A feed point should be provided ahead of the primary clarifier for feeding of polymers and of primary oxidants, such as chlorine, hydrogen peroxide and potassium permanganate. At the time of the concept design, it may not be evident that chemical feed actually will be needed. Therefore, the feeding and storage equipment need not be installed initially, but the piping and valving at the feed point should be in place, and adequate space should be provided for feed and storage equipment. The ability to feed chemicals ahead of the primary clarifier will permit the correction of the following annoying problems:

- oxidation of detrimental wastewater components such as sulfides and some industrial chemicals
- capture of unusual colloidal solids in the primary clarifier

Nutrient Feeds

A feed point should be provided in the piping between the primary clarifier and the aeration basin for the feeding of nutrients. Use of the feed system would become necessary if industrial wastewaters contributed an organic load not containing nitrogen and phosphorus and resulting in a nutrient demand greater than that available in the combined domestic sewage. The preferred sources of nutrients generally are aqueous ammonia for nitrogen and phosphoric acid for phosphorus. Again, if the immediate need is not obvious, provide the space for future installation of feeding and storage facilities.

When the organic and nutrient ratio is variable during a 24-hr period, the generally accepted average phosphorus:nitrogen:BOD_5 ratio of 1:5:100 is not an adequate guideline for optimum biomass performance. Information for forward feed control of nutrients is not available, and shortages of nutrients for a few hours during rapid synthesis is detrimental to the biomass characteristics. During the periods of rapid synthesis, an excess feed of nutrients may be necessary.

Returned Sludge

Feeding of chemicals to the returned sludge usually is done to control unwanted filamentous organisms. Common choices of chemicals are chlorine and hydrogen peroxide. Chlorine is effective in controlling most bacterial and fungal filamentous organisms. When chlorine is used, the effluent quality will be affected by a discharge of colloidal solids. Alternatively, hydrogen peroxide can control filamentous bacteria by attacking the organism's sheath. While this chemical feed approach has little impact on effluent quality, the use of hydrogen peroxide is more expensive. There is little doubt that problems with filamentous organism will be encountered. Therefore, the facilities for feeding of chlorine and hydrogen peroxide to the returned sludge should be provided in the initial design.

Secondary Clarification

Chemical feeds to the effluent at the process flow between the aeration basin and the primary clarifier are made primarily for better capture of solids by the clarifier. Polymers are employed under emergency conditions. The performance of polymers is improved if flocculation is provided before settling. Therefore, the secondary clarifier should be designed as a clariflocculator, or a flocculation unit should be placed between the aeration basin and the clarifier. When the suspended solids requirements of the effluents are relatively stringent, roughly 15 to 20 mg/l, the facilities for polymer feed and for flocculation should be included in the initial design. Flocculation is optional where the effluent requirements are less stringent, but a feed point for the addition of polymer definitely should be provided.

PIPING

In the design of piping, there are very important criteria which frequently are not given adequate consideration during detailed design.
1. Buried piping is difficult to maintain and repair. In general, all piping, but most definitely all critical piping for forward flow, sludge handling, and chemical feed, should be placed in pipe galleries.
2. The suction piping absolutely must be as short as possible wherever pumps are used to transfer sludge or scum. Limit suction piping for scum to one pipe diameter, if possible.
3. Chemical feed lines, particularly chlorine feed lines, must be as short as possible. If long lines must be considered for delivery of metered flow to several locations from a single storage area, run these feed lines at locations which are accessible for service and inspection.
4. In the sludge management system, the design may encompass single units. If so, bypass lines must be provided so that sludge may be taken from the system when the single unit in the process scheme is out of service.
5. Sampling lines must be kept short. If the sampling line is too long, the line becomes a biological contact unit, thereby changing the nature of the effluent sample significantly.

DESIGN OF FACILITIES FOR OPERABILITY

6. The piping system should include a drain system which will permit any tank in the facility to be drained completely.

7. When a suction lift is involved, progressive-cavity pumps will not operate on sludge successfully. The potential for suction plugging is too great, and if flow is stopped the pump stator will be destroyed.

8. The inlet piping to clarifiers must not involve a high head differential for the flow from the preceding unit. Too much kinetic energy in the center well can generate currents in the clarifier. The head differential should be kept low, or a design to dissipate the kinetic energy should be introduced.

SAMPLING POINTS AND FACILITIES

Major sampling points are those locations involved in the results reported to regulatory agencies and in recording standard process performance. In addition, there are in-process sampling points that are needed for operational information.

Raw Wastewater

From a regulatory point of view, a raw wastewater sample represents the facility's influent flow before processing and ahead of any returned flows within the plant. In most wastewater treatment plants, the raw wastewater is pumped directly into the facility, preceded only by screening. Side streams within the process, such as supernatants, filtrates and tank drainings, may return conveniently by gravity to the wet well at the raw wastewater pump station. Where side streams are handled in this manner, a sampling location for raw wastewater must be provided between the bar screen and the raw wastewater wet well. Because of larger particle sizes present in the wastewater, the collection of a representative sample at this location is difficult. The sample should be transferred by a chopper-type pump that will reduce particle size and improve the performance of the sampler.

An alternative arrangement would be to sample after comminution. This approach requires a separate pump station for side streams so that they may be returned to the inlet of the primary clarifier after the sampling point.

Final Effluent

Sampling of the final effluent is significantly less complicated, but even here, the importance of a short sample line should be apparent.

In-Process Sampling Points

The in-process sampling points are those between each major unit operation and on all side streams and sludge streams. For the most part, these sampling needs are met by pipe taps which give access to process flows. For reasonably representative sampling of solids, the pipe taps should be at locations of turbulence in the pipe, such as pump discharges and elbows.

INSTRUMENTATION

Flow Monitoring

For the same reasons that raw wastewater samples are difficult to obtain, it may also be difficult to get a raw wastewater flow measurement without including returned flows. Generally, measurement of the forward flow at the effluent will be adequate for most flow monitoring purposes. It is important, however, to provide flow measurement for all sludge streams and recycle streams within the process.

pH Monitoring

Raw wastewater should be monitored for changes in pH so that action may be taken if potential process upsets are encountered. Biological activity also may bring about changes in pH, such as increases in pH when the salts of organic acids are oxidized or decreases in pH when ammonium salts are oxidized. Monitoring for pH should be conducted at the influent and effluent of the aeration basin.

Process Control Instrumentation

Operations personnel need clear information about the status of the kinetics within the activated sludge process. The operators may not understand kinetics as a term, but they can understand changes in meaningful measurement that will have a direct impact on the operation of the process.

Dissolved Oxygen

Dissolved oxygen (DO) concentrations in the aeration basin readily are interpreted as either too low for the process or too high. If DO concentrations are too high, power is wasted. Therefore, it is worthwhile to provide continuous monitoring of mixed liquor DO. For larger systems, automated modulation of oxygen transfer based on DO measurement is feasible.

Oxygen Uptake Rate

The potential impact of the wastewater on the biomass and the process can be interpreted directly by oxygen uptake rate determinations made on mixtures of raw wastewater and returned sludge. An uptake rate less than the rate of the returned sludge alone means inhibition or toxicity. The uptake rates above the uptake level of the sludge correlate with strength and/or type of incoming organic load. With experience, uptake rate measurements can become an indispensable tool for the operator. In those systems where the industrial load is significant, automated uptake rate instrumentation should be considered. In other systems, manual instrumentation should be provided for use by the operator.

Solids Measurements

The operator needs information on the concentration of suspended solids in the returned sludge and the mixed liquor. Among the instruments available, those based on light transmission appear to provide a reasonable approach which could furnish useful information, but it seems that none of the instrumentation currently on the market has suitable dependability. Engineers should continue to seek suitable instrumentation for suspended solids measurement.

Clarifier Sludge Blanket Level

Knowledge of sludge levels in the secondary clarifier is important for sludge management. Under normal operating conditions, the level of sludge is kept to just a few inches. Changes in the level of sludge due to changes in pumping rates or sludge settling rates should be brought to the operator's attention quickly. Sonic instrumentation is available which can monitor the blanket level; this equipment should be considered in the concept design. For manual monitoring, a Sludge Judge—a clear plastic tube with a foot valve—may be employed. With the Sludge Judge, a sample representing a top-to-bottom core is collected, thereby providing excellent information on the supernatant and sludge layers.

Operations Laboratory

As a matter of course, each wastewater treatment facility must have a main analytical laboratory which conducts those analyses required for recording and reporting the performance of the facility. In addition, there should be an operations laboratory where the operations personnel can make determinations that are useful in operating the process immediately. Useful measurements in operations work include:

1. dissolved oxygen and oxygen uptake rate measurements for monitoring of oxygen transfer to the mixed liquor, the biomass status, and the organic load to the process
2. a pH instrument suitable for use with specific ion probes so that selected measurements of ammonia, nitrate or other ions can be determined if necessary
3. a settling apparatus suitable for determination of initial settling velocities and sludge compaction curves for estimation of clarifier performance
4. a microwave oven or other technique for a quick determination of suspended solids concentrations
5. a microscope for inspection of the biomass for type and toxicity of biological organisms

The convenient use of these aids will put the operator in reasonable contact with the process.

SLUDGE MANAGEMENT

Thickening

Thickening is a necessary operation for almost any subsequent processing of sludge. Primary sludges thicken readily and generally are handled suitably by gravity thickening. Waste activated sludges do not respond readily to gravity thickening; better thickening is achieved by air flotation.

Storage

Liquid Storage

For further processing, separate storage of primary and secondary sludges should be practiced generally. For purposes of dewatering of nonstabilized sludges, blending can be impractical unless this activity is done carefully and completely. Variations in the blend would cause variations in the conditioning requirements and result in an operational headache. Careful blending also is important before incineration to achieve relatively uniform Btu values. The storage basin should be mixed, covered, and fitted with odor control for the off-gases, and there should be adequate storage so that the activites of sludge management—including downtime for equipment—can be scheduled.

Cake Storage

Incineration operations usually are designed for combined operation of dewatering and incineration facilities. Limitations in the performance of either operation result in a reduced rate of sludge processing. Additional flexibility would be introduced by cake storage. Cake storage requires bins with breakers to move the sludge. Sludge at 23 to 25% solids has been pumped successfully with a progressive-cavity pump. The possible benefits of cake storage should be considered further.

Oil Reclamation

With the current price of oil, the skimmings from the primary clarifier may represent a significant source of energy for a wastewater treatment facility. By following standard oil reclamation techniques, such as acid cracking and filtration, the reclaimed oil has about the same characteristics as a Number 2 Fuel Oil. In one instance, a large wastewater treatment facility that incinerates sludges hopes to provide all needs for supplementary energy by using oil recovered from primary skimmings.

CONCLUSIONS

From the operations point of view, the more commonly seen approaches to the design of activated sludge facilities, particularly those involving smaller facilities, have limited flexibility. The lack of flexibility is related primarily to decisions based on savings in capital costs. A basic design flaw from the operations point of view is the provision of a limited number of parallel units, which significantly limits the process and causes a reduced quality of effluent when one unit of an operation is out of service.

The fact that limited flexibility exists is partially recognized by the engineer, but since the designer generally has limited experience in operating a facility, the total degree of the limitation is not understood, and the engineer finds the design of more process flexibility too costly.

During review of the process design, the process limitations are given tacit approval by the regulatory agency, but once the facility is in operation, the enforcement branch of the regulatory agency expects the process to meet effluent requirements, regardless of maintenance problems.

In general, the operator's view of an activated sludge design that has good operability and maintainability is one that encompasses the following design features:

1. a number of parallel units in each unit operation, which permits one unit to be out of service without significantly affecting the performance of the process
2. pipe galleries for easy inspection and maintenance of all important piping in the facility
3. an operations laboratory for use in the operations effort
4. the opportunity to make process modifications readily available and to adjust to emergency conditions in the process piping
5. the opportunity to use chemical addition throughout the facility

Proposed flexibility in design adds to capital costs. The amount is probably 25 to 40%, with a higher amount inversely related to smaller sized facilities. Though the value of this additional investment is difficult to assess because much of the benefit is derived from the abatement of operational nuisances, it seems very probable that the return is well worth the initial expense.

CHAPTER 10

GOOD PROCESS CONTROL: PLANNING AND IMPLEMENTATION

Ed R. Fioroni and Walter D. Frais
   Polysar Limited
   Sarnia, Ontario, Canada

INTRODUCTION

In December 1976, the Ontario Ministry of the Environment issued a Ministerial Order to Polysar requiring a reduction in effluent loadings. On March 5, 1980, the Ministry indicated provisional acceptance of a proposal for a biological oxidation treatment system. At that time, Polysar proposed to construct an extended aeration, activated sludge process unit with suitable pretreatment, for the treatment of water-soluble organic compounds by biological oxidation. This proposal met with the conditions of the "Requirement and Direction" by reducing the total organic carbon (TOC) of treated wastewater by 50% and the phenol by 75%.

At this point, Polysar management, after thorough investigation, decided that it was of key importance to focus on what appeared to be the three major historical problems in other North American activated sludge plants:

1. proper design
2. instrumentation (on-line and laboratory)
3. pre- and post-startup training of personnel

To ensure that these three requirements would be fully satisfied, an experienced waste treatment design engineer was given the responsibility to properly design the bioxidation plant to allow for process *flexibility* and *controllability*.

It was also clearly understood and highlighted early in the game that the single most important factor for a successful operation was going to be the *process operator*. It was important to perceive what a wastewater treatment operator does and recognize that years of experience supplemented by adequate training was needed to master the essential operator job knowledge and skills. From this, a solid and effective pre- and post-startup program had to be established to ensure that the operator could learn to properly interpret process information and apply it correctly to a flexible-design plant. Let us look at the three-part plan used by Polysar.

PART 1—DESIGN OF THE PLANT

Polysar, realizing that future process modifications and expanded or new production facilities could result in higher hydraulic and/or pollutant loadings, decided to construct an extended aeration biological oxidation plant with a high degree of process flexibility. For this reason, the biox plant was designed to handle in excess of 200% of the average daily hydraulic flow and TOC loading.

The plant consists of a pretreatment section followed by a biological section. The pretreatment section (Figure 1) consists of pH adjustment, oil and suspended solids separation and removal and preaeration/equalization.

Contaminated wastewater is pumped continuously to the biox plant from three separate sources within the boundaries of the overall plant. The major wastewater stream originates at the copolymer effluent treatment system which receives wastewater from three separate synthetic rubber and latex production facilities. This stream passes through dissolved air flotation (DAF) unit, followed by sand/anthracite filters for suspended solids removal, before being pumped more than 0.6 mi to the biox plant.

The second wastewater stream originates at the main oil separator, which receives oil-contaminated water from up to 11 different production areas. Free oil is separated by gravity and the resultant water is then pumped 0.5 mi to the biox plant. The third stream originates at the isobutylene extraction production unit. The three streams are well mixed in the influent shaft of the biox plant.

The stream then passes through a two-stage pH control system (15 min residence time). Each pH tank is equipped with a 4-speed agitator to

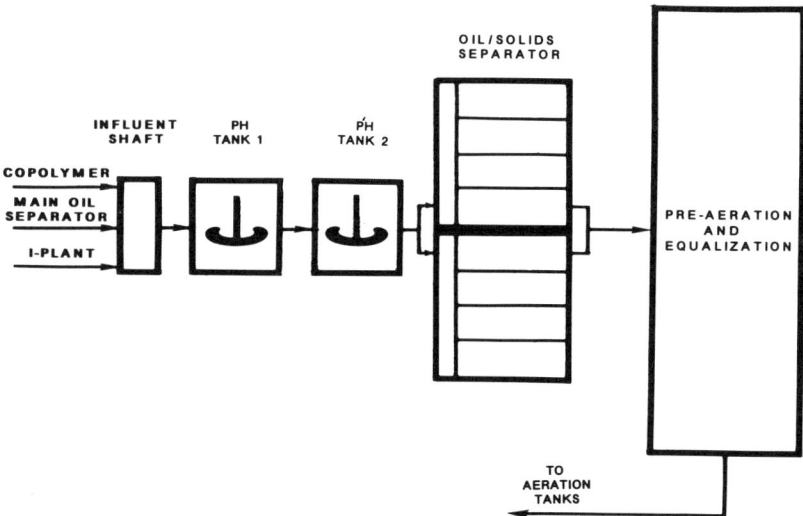

Figure 1. Pre-treatment section.

supply the required degree of mixing. Sulfuric acid (93%) and caustic (25%) are the chemicals used. Facilities have been supplied to add aluminum sulfate, ferric chloride and/or up to three different polyelectrolytes into either of the pH tanks for oil or solids removal in the API separators. Since the influent streams are deficient in phosphorous and nitrogen (nutrients), facilities to meter phosphoric acid (75%) and ammonium hydroxide (28%) have been supplied. The pH-adjusted stream then passes through two banks of four API gravity separators in parallel. One bank of four can be removed at any time for maintenance.

The last stage of the pretreatment section is preaeration/equalization. This tank is 40 × 8 m (130 × 26 ft) with a variable depth of 7.5–10 m (25–33 ft). It is equipped with one jet aeration distribution header to supply the required mixing and oxygen transfer. In terms of equalization, the tank is well mixed, supplying a hydraulic retention time of 2.8 ± 0.4 hr at design flows. The flow to the aeration tanks is smoothed by means of a control valve and tank level fluctuation. The jet aeration system also allows for preaeration of inorganic chemical oxygen demand (COD) such as sodium bisulphite and volatile organic COD. The overall objective is to maintain a minimum dissolved oxygen (DO) concentration of 2 mg/l in the feed to the aeration tanks. The nutrients can also be added at the discharge of the preaeration/equalization tank.

The pretreated water stream then enters the biological section of the plant consisting of four aeration tanks and four clarifiers. Each of the four aeration tanks holds 6000 m^3 (1.5 million gal) and is equipped with three jet aeration distribution headers. This makes each tank a completely mixed unit. The tanks are 10 m (33 ft) in depth, having been designed for optimal oxygen transfer and energy conservation. With the cold Canadian winter climate, it is critical to keep the energy in the system. This part of the process has been designed for maximum flexibility. The aeration tanks can be used in any one of the following three configurations (Figure 2):

1. one, two, three or four aeration tanks in parallel
2. one or two chains of two aeration tanks in series
3. one chain of up to four aeration tanks in series

Configuration 1 will be used for the startup and thus will be further discussed.

In the parallel configuration, a central feed water distribution chamber controls the flow to each of the four aeration tanks. A similar central distribution chamber allows complete control of how much return sludge is sent to each of the aeration tanks. Based on design flowrates, hydraulic retention times of 6, 12, 18 or 24 hr can be achieved by using one, two, three or four of the aeration tanks. Each aeration tank, as well as the preaeration/equalization tank, is equipped with a defoamer/water spray system to control excessive foaming.

The aeration system consists of three variable-rate air blowers, operating in parallel, feeding three jet aeration distribution headers per aeration tank and one jet aeration distribution header in the preaeration/equalization tank. During maximum oxygen demand, two of the air blowers will supply sufficient dissolved oxygen, with the third blower on standby. The aeration system has been designed to provide a maximum of 36,000 kg/day (80,000 lb/day) of dissolved oxygen.

The overflow from the aeration tanks is fed through a common line to up to four conventional gravity settling clarifiers in parallel for separation of biological solids (Figure 3). A static in-line mixer has been installed in this overflow line. Addition of aluminum sulphate, ferric chloride or up to three polyelectrolytes before and after the in-line static mixer, has been provided to assist, if necessary, in the settling characteristics of the biological sludge.

The clarification system consists of four clarifiers, each 28 m (92 ft) in diameter, with a 3.4-m (11.2-ft) sidewall depth. A center feed, peripheral overflow system was designed. A surface skimmer system has been provided for each clarifier to permit collection of scum, floating oil and solids.

PLANNING AND IMPLEMENTATION 157

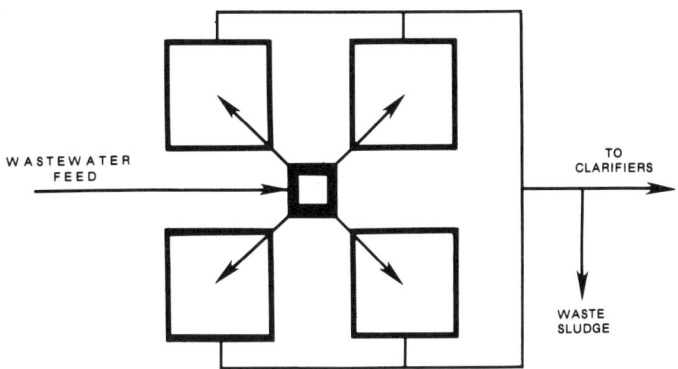

(1) **FOUR AERATION TANKS IN PARALLEL**

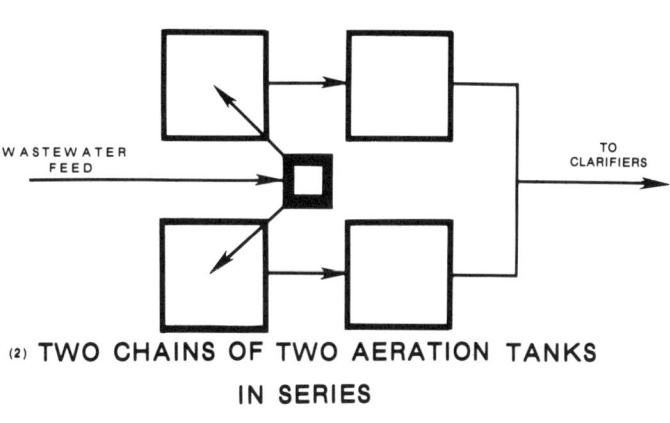

(2) **TWO CHAINS OF TWO AERATION TANKS IN SERIES**

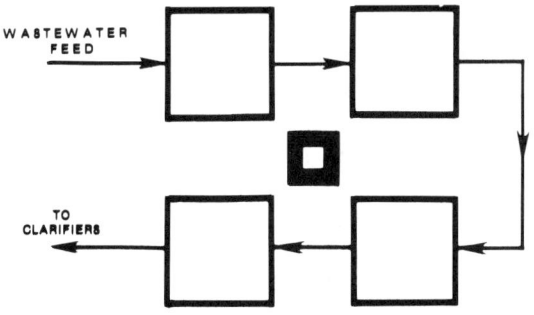

(3) **FOUR AERATION TANKS IN SERIES**

Figure 2. Aeration tank configuration options.

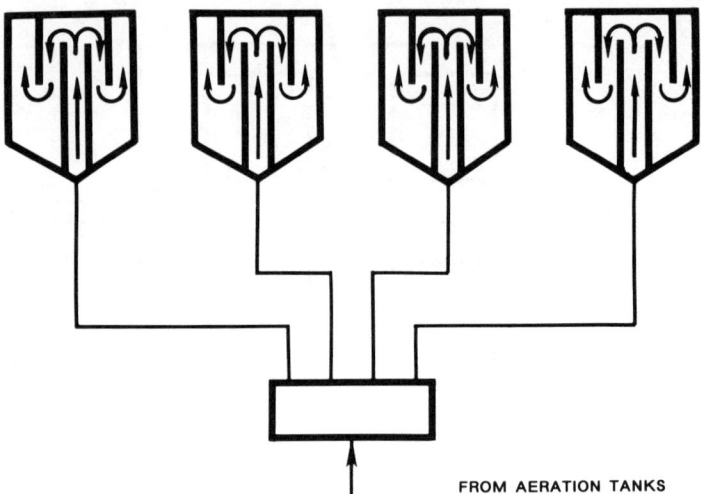

Figure 3. Clarifiers.

Settled sludge from the clarifiers is returned to the aeration tanks by means of variable-speed pumps. The sludge return system has been designed to provide a recycle flow capacity ranging from 50 to 200% of the average design flow. The maximum recycle flow is provided by two variable-speed pumps, with an additional unit on standby. Each pump recycles settled sludge from two clarifiers which discharge into a common line back to the aeration tanks. The clarified treated effluent water is discharged to the St. Clair River via an extended outfall.

Figure 4 depicts the physical layout of the biox plant.

PART 11—INSTRUMENTATION (ON-LINE AND LABORATORY)

A properly designed biox plant which allows for process flexibility and controllability is useless unless the operator has the required process information readily available to judge instantaneously how effectively the process is functioning. On designing the instrumentation package for the plant, the following four factors were considered;

1. Eleven different operating units contribute to the wastewater to the biox plant.
2. Over 400 different chemicals enter the plant as raw material. Any of them in their pure form or in any combination could find its way to the biox plant.

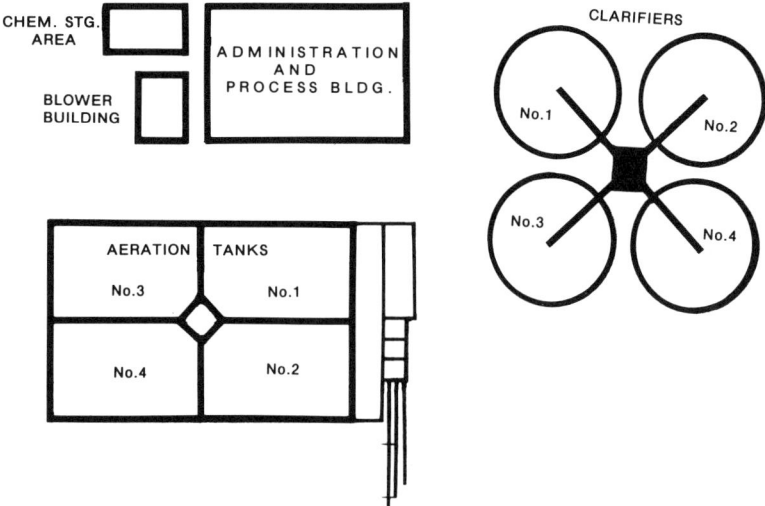

Figure 4. Biox plant.

3. Some of the 400 different chemicals in certain concentrations could prove toxic to the biological sludge.
4. Pollution loadings in the feed streams, based on historical TOC and COD data, showed high daily and/or hourly variability.

Based on the above four factors, the instrumentation package was designed to:

1. warn the operator immediately of an unfavorable condition that could prove toxic to the microorganisms
2. give the operator immediate information as to the health of the system so he could take effective action
3. supply the operator with historical data that could result in process improvements

The instrumentation package was therefore designed as described in the following paragraphs.

Each of the three influent feed streams is equipped with on-line instruments to continuously measure flow, pH, temperature and oxydation reduction potential (ORP). To maintain the pH between 6.5 and 8.5 at all times, a two-stage feed forward pH control system is employed. In the first tank, based on two independent continuous pH measurements, the computer controls the amount of sulfuric acid or caustic added to roughly bring the pH into the desired range. The second tank is for fine tuning of pH, again being computer-controlled based on two separate pH probe measurements.

At the discharge of the oil/solids separator, after the pH has been corrected and any free oil or solids have been removed, two key instruments have been placed on-line to protect the biological portion of the biox plant from potentially toxic or inhibitory components. The first instrument is an ASTRO on-line TOC analyzer. This will warn the operator immediately if either an extremely high or low TOC loading is entering the system. Although TOC does not indicate whether the material entering the plant is toxic, inhibitory or a good food source, it does warn the operator immediately that a deviation has occurred, and that further investigation or testing is required. Whether the incoming material is toxic, inhibitory or a good food source is determined by the second on-line analyzer, the Arthur Respirometer. The on-line respirometer takes the correct proportion of feed and return sludge to form a mixed liquor volatile suspended solids (MLVSS) concentration to simulate conditions in the aeration tank.

After aerating the sample continuously in a closed system, the oxygen consumption rate is measured; this rate can be related directly to the breathing rate of the microorganisms. This is a 15- to 30-min batch test. In effect, this instrument tells the operator *now* what would happen in the aeration tank in 3-4 hr if nothing were done. Although this respirometer could be used in this location to measure directly the food entering the system, in our case, this instrument's main function is to detect a *toxic* material. This will give the operator a chance to react (3-4 hr) before the toxic material would enter the biological portion of the plant.

Polysar feels that the combination of the on-line TOC analyzer and respirometer will give sufficient warning of a toxic shock load to the system. Lines to both instruments will also allow the sampling of the stream leaving the preaeration/equalization tank, should the need arise.

The equalization/preaeration tank is equipped with a temperature probe and two DO probes. The operator will have to manually adjust the valve on the air line to maintain the DO in the tank at the desired level. To smooth out the water flow to the aeration tanks as much as possible, a control valve tied into a level controller on the tank will allow the tank level to fluctuate ± 2.5 m (8.2 ft). This function is handled by the computer.

The aeration tanks are controlled on the basis of constant DO concentration. Each aeration tank is equipped with two on-line DO probes.

In total, the eight DO probes send a signal to the computer, which automatically averages the observations and compares this value to the DO set point specified. The computer then makes the necessary adjustment to the air discharge from the blowers. A joint project with the Water Technology Center (WTC) in Burlington, Ontario, Canada resulted in the formation of an oxygen consumption optimization algorithm. This algorithm will be incorporated in the DO control function of the computer.

In addition, each aeration tank is equipped with a temperature and pH probe and feed water flow measurement. A second on-line respirometer, along with a sample sequencer, will be used to measure the respiration rate of the microorganisms in any of the four aeration tanks, as well as in the return sludge. Sludge can be wasted from either the discharge of the aeration tank or the return sludge line. In both cases, the computer controls the flow of waste sludge via a flow control valve.

The clarifiers are controlled by sludge level. A sludge level detection strip on the side of two of the clarifiers determines the sludge/water interface. Level controllers on the two clarifiers sense the interface and send the message to the computer. The computer compares these readings with the set points and adjusts the motor speed on the two variable-speed pumps, thus pumping more or less sludge back to the aeration tanks. The second clarifier in each couple goes along for the ride. Each clarifier is equipped with a high sludge level alarm.

The return sludge line to the aeration tanks is equipped with a DO, ORP, pH and turbidity probe, flow measurement and, on a sequenced basis, the second respirometer. The clarified treated effluent water discharge line is equipped with a DO, ORP, turbidity, pH and temperature probe. The flow is also measured.

Composite samplers on the three feed streams, the preaeration/equalization tank influent and effluent and the clarifier effluent give all the information required to evaluate the effectiveness of each stage of the treatment system. An additional composite sampler on the final effluent line outside the unit boundary allows Polysar to independently monitor the biox plant.

The computer, besides all the functions mentioned, is also a data acquisition system. All on-line instruments mentioned send signals back to the computer where the data are stored, averaged and logged.

Besides all the on-line instrumentation, the plant has a modern, well equipped laboratory. On a continuous basis, the following 12 streams are pumped into the control laboratory, where they can be observed and sampled:

- the three feed streams
- after API separator
- after preaeration/equalization
- each aeration tank
- combined aeration tank overflow
- clarifier overflow
- return sludge

The control laboratory has been equipped to carry out the following tests:

- TOC (ASTRO)
- BOD_5
- phenolics (Technicon*)
- ammonia (Technicon*)
- pH (Orion)
- ORP (Rexnord)
- DO (Rexnord, YSI)
- jar testing
- settling tests (Arthur Settleometer)
- phosphorous (Spec. 21)
- MLSS, SS
- MLVSS, VSS
- plate counts
- titration
- solvent extraction
- turbidity (Hatch)

In addition to the above routine tests, the following key tests are utilized on a daily basis for both process control and research purposes:

1. microscope
2. adenosine triphosphate (ATP) (Turner)
3. zeta potential (Zetameter Inc.)
4. two lab-scale respirometers with automatic vents (Arthur Technology)

In addition, two portable pilot plant units have been constructed to be located in the lab which will allow us to experiment on a continuous basis with any potential process modifications. The laboratory also includes:

- sterilizer
- climate-controlled walk-in incubator
- five positive airflow fume hoods
- Barnstead water still
- Barnstead ultrapure water system

---

*The Technicons are equipped with microcomputers.

PART III—TRAINING PHILOSOPHY

The capability to train operators exists. The questions that were difficult to answer at the start were: Can we educate people to take advantage of the training opportunities available and, most of all, can we teach upper management what an operator is? And, if we are able to teach them, will they believe us? (It is well known that strange ideas prevail in the outside world that wastewater operators are not required to have broad and technical knowledge.)

Having realized that we were dealing with a powerful union, our first obstacle was to convince the union that we required experienced and skilled operators in the wastewater field. Our second objective was to communicate this information in a favorable manner to upper management. At the beginning, it was very difficult for them to understand the validity of our convictions. They could not visualize "waste" requiring individuals that needed extensive experience and adequate training.

Looking back, the task of educating people not involved in the process itself was the most difficult hurdle to overcome. How did we achieve this goal? We gave some very sad examples of how major bioxidation plants were unable to meet the desired requirements because of lack of proper pre- and post-startup training.

After long and difficult communication meetings, the authorization was given by upper management and the union to come up with an extensive and satisfactory training program. At this point, we divided our training requirements into three parts:

1. pre-startup training
2. startup training
3. post-startup training

PRE-STARTUP TRAINING REQUIREMENTS

It was decided that the biox plant would require six operators, six operator assistants, two maintenance people, a process engineer, a foreman, a secretary and a plant manager. The following was the timetable suggested:

1.	plant manager	onsite	2 years in advance
2.	process engineer	onsite	18 months in advance
3.	foreman	onsite	15 months in advance
4.	operators	onsite	12 months in advance

1. We initially recommended that all six operators be hired from outside Polysar; this would give us a solid baseline experience and knowledge. Due to union pressure, we were only able to obtain three outside operators. These three individuals all possessed university degrees and 6-20 years of wastewater experience, and were very *well versed* in activated sludge problems and startup procedures.

2. Our next step was to hire the remainder of the required operators and assistants within Polysar. We decided to request the top individuals that were interested in working in an activated sludge plant. The interest was very high and over 100 persons applied for the job. That made it much easier to obtain top operating staff within the plant.

3. One of the problems associated with wastewater treatment plants is that operating personnel do not stay on the job for very long, causing discontinuity and lack of knowledge derived from minimal onsite experience. We had to develop a sound, progressive training program to ensure that our operators would consider their job a permanent one.

4. The decision was made that the only way the operators could leave the biox plant was if they personally chose to post out. This decision enabled us to ensure the *permanency* of the position of "Biox Operator" and allowed us to design a solid progressive certification program.

5. In Polysar, laboratory technicians have a different job description than process operators, and one cannot perform the other's duties. Therefore, we had to design a job description that would fit both categories and at the same time be accepted by the union. We did this by convincing management and union that the operator will be efficient, if he personally performs the required tests and does the actual interpretation. This dual responsibility will make his job more enjoyable.

6. Since process control will be achieved by a data acquisition and control system (DACS) which is to include a central operator's console and process interface cathode ray tube (CRT) units, the operators had to be properly and thoroughly trained to operate and understand the computer system.

A training program consisting of three full days of theory and three full days of practical application was given four months prior to the targeted startup date. This program would help the operator to:

a. understand the console operation and properly interpret data through process control
b. minimize the sense of mystique and terror that the monstrous but simple machine gave to individuals that were not properly trained

Every operator, assistant operator, foreman, engineer, secretary, and all instrument people and maintenance personnel took the course. Although the course was an excellent tool, we strongly recommend that this type of training be carried out during or just before actual startup. Our course will be repeated three weeks before startup.

7. At this stage, we had permission to hire three outside operators and to obtain the top candidates from within the plant. We established a permanent position for our personnel and were successful in outlining the responsibility of the operator's within their job description. One key item was still missing: we had to convince Management that we required these individuals at least *six months to one year before the plant actually was operational.* During this time frame, we would ensure proper operation and maintenance of the wastewater system, in accordance with the design, to achieve maximum plant performance and efficiencies of treatment. We established certain requirements necessary for a professional and effective training program. The trainer would be able to:

  a. correctly define and categorize problems and determine whether they can be solved through training
  b. analyze the tasks performed at the plant to determine the training needs
  c. determine the type and level of knowledge required for each task
  d. specify instructional objectives that will be taught to the employees
  e. develop evaluation activities to determine the progress of students and the effectiveness of the training itself
  f. choose the best method of instruction for specific objectives
  g. choose and create effective media for different types of levels of objectives
  h. communicate with other people on the job site
  i. deliver a lesson effectively
  j. revise and evaluate training materials to remain current in a dynamic profession
  k. develop a training schedule

8. The first six months would be dedicated to achieving the above; the next six months would require training the remaining operating personnel. A training schedule was prepared. There was one major concern: nine out of the twelve operators had no biox experience. We decided to install a major pilot plant so that our operators could use it as a training and learning tool. Two identical separate systems were built. Each system consisted of:

  a. one 400-liter feed tank (with agitator)
  b. one 78-liter aeration tank (The aeration system was under continuous agitation with airflow control.)
  c. one 21-liter clarifier with rotating mechanisms (1-5 rpm)
  d. two peristaltic pumps (0-500 ml/min) for feed and return sludge

With the aid of the pilot plant, the operators learned how to define a problem and find its possible solution. They gained expertise in operating instruments and laboratory machinery, but most of all they were able to understand that an activated sludge plant will offer many problems that will require interpretation gained by a solid technological approach (i.e., actual microscopic analysis and interpretation). The operators very quickly learned *not to overreact*.

9. Our last pre-startup objective was to send the operators to an operating activated sludge plant within our area for a few days to give them some practical experience. Every operator was sent to an activated sludge plant for three days, with very positive results. It was at this time that we realized how effective our training up to that point had been. Our operators had more basic knowledge than most of the other operators to whom they talked. This was primarily because we were able to obtain our personnel *1 year before startup* but also because we had made our job description *a permanent one*. From this, we made a serious commitment to ensure our operator's continuity and job security within the biox plant.

Summarizing, let us analyze the pre-startup achievements:

a. defining the basic need for manpower
b. obtaining personnel with practical experience and vast wastewater knowledge
c. instituting a security in the job requirement to ensure that the operators will not move
d. ensuring that the job description included a clause for operators to have dual responsibilities—laboratory and process
e. developing a training schedule
f. obtaining all operating personnel 6 months to 1 year before actual startup
g. constructing and operating a miniature pilot plant
h. visiting other bioxidation plants within a specified area
i. training

STARTUP TRAINING REQUIREMENTS

The startup procedure, at this point, was well established through the use of the pilot plant. The operators, with the aid of the pilot plant, were able to actually experiment and collect data in determining which method would be most advantageous. After analyzing all the various facets of our needs, we decided that our operators should be involved in overseeing the final phase of the construction. This would satisfy two major requirements:

1. It would give the operators hands-on experience of all equipment (they would actually see the internals), lines, laboratory facilities and pumps. By comparing in detail the drawings with the actual construction, the operators would be able to point out outstanding errors and possibly have an early input in design changes. In short, they would be able to learn how the plant physically operates.

PLANNING AND IMPLEMENTATION 167

2. It would ensure that all design and construction parameters were met satisfactorily before the plant became the direct responsibility of the biox team.

We assigned responsibilities during this period as follows:

Operator 1 clarifier area
Operator 2 underground and overhead piping and pipe gallery
Operator 3 laboratory and building
Operator 4 mechanical areas and chemical area
Operator 5 aeration and preaeration
Operator 6 pretreatment, oil separation and instrument room

We also had to train the plant and its various units to understand what a biox unit is and what its purpose is within the overall plant structure. We initiated short, sound, professional presentations outlining the plant design and trying to explain what the consequences could be if effluent conditions were not met. Our objective was to instill in our units a sense of appreciation for the high technology required and a devotion needed within biox, but also to ensure a sense of respect and loyalty to our plant. We had to radically change ideas and conceptions that are vividly exemplified by the common statement: "Don't worry, send it to the biox plant, they will take care of it." These presentations lasted 30 min and were followed by 1-hr discussions repeated four to five times during the first year of operation.

POST-STARTUP TRAINING FORMAT

To minimize labor/management relations, it was mandatory to outline a certification program and make it part of the job requirement. Certification ensures the following:

1. It will allow operators to maximize their potential.
2. It will ensure that the operators will perform with the most innovative and modern technology.
3. It will motivate the operators to continuously learn new technology and operation strategy, and will also enable them to more clearly understand the intricate problems associated with the biox plant.
4. It will ensure an overall "good feeling" amongst the entire crew arising from self-confidence and ability.
5. It will save Polysar a tremendous amount of dollars:
    a. dollars derived from a better DO control (less electrical cost)
    b. dollars derived from an overall improved control (less maintenance cost)
    c. dollars derived from an ability to interpret results and make necessary changes that will avoid major problems
    d. dollars derived from an overall better environment and increased work production

The *certification* of operators will be based on *experience*, *education* and *examination*, with provision for limited substitution of education and experience with each other under an approved system. Formal education can be supplemented by approved training courses, seminars, etc., for which continuing education units (CEU) are allocated.

Considering the fact that in Ontario, there is no mandatory or voluntary certification program as of yet, it is noteworthy that we at Polysar have developed a four-phase training program that will eventually lead to a complete certification when it is established in this province. Let us look at the four-phase training program developed.

Phase 1

This phase is intended to give the staff a review of grade 12 and first-year college of the subjects that they will require in the everyday operation of the plant. The subjects are:

- mathematics   16 units
- chemistry   8 units
- biology   2 units
- physics   2 units
- microbiology   3 units

Each operator will be writing a pre-test on each subject. The result of the test will enable us to place them in the appropriate level. Some operators will not require this first phase and will automatically go to phase II. This will be all in-plant training. To institute an effective in-plant training program, existing technical personnel were utilized in development and delivery. We expect an operator to complete phase 1 within 3 to 6 months.

Phase II

During this phase, they will be taking a basic laboratory course, developed and implemented within the plant. They will also start having hands-on experience through the digestion and understanding of the operation and maintenance manuals. The objective of this phase is to introduce them to the laboratory technology (theory and practice) and also allow them to slowly digest the operation of the plant via a practical approach. This should take approximately 6 months to 1 year.

Phase III

During this phase, the operators will be concentrating on an advanced laboratory course in which they will learn to accurately interpret results and make necessary changes. During this phase, they will attend the "application of on-line respirometry and other analytical instrumentation process control course." We feel that this course is imperative since our operational strategy is based on respirometry and DO control. In this course, they will be exposed to a practical approach to understanding the science of respirometry and settling properties. They will then attend a 60-hr wastewater course at our community college. This course is a prerequisite for level IV. This phase should take approximately 1 year to 18 months.

Phase IV

On completion of Phase III, the operator is ready to undertake the Ministry Courses offered in Toronto. We have chosen eight courses to be taken over a period of three years. (Table I). At this stage, the operator will possess the experience supplemented by adequate training to be considered a "qualified wastewater operator." This will be to his advantage when certification becomes compulsory.

Summary of the Training Program

Phase I

Basic Knowledge:

- mathematics   16 units
- chemistry   8 units
- biology   2 units
- physics   2 units
- microbiology   3 units
    up to 6 months

Phase II

- introductory laboratory course
- O&M exposure with hands-on experience
    6 months to 1 year

Table I. Ministry of Environment Courses

Course	Experience	Education	CEU	Prerequisites
Basic Sewage Treatment Operation	6 months	Lambton College	3.0	Lambton College
Activated Sludge Workshop	1 year	Lambton College	3.0	Course No. 1 + Lambton College
Primary Treatment and Digestion Workshop	1 year	Lambton College	3.0	Course No. 1 + Lambton College
Basic Water Treatment Operation	1 year	Lambton College	3.0	Lambton College
Surface Water Treatment	18 months	Lambton College	3.0	Course No. 1 + Lambton College
Pump Operation Workshop	1 year	Lambton College	3.0	Lambton College
Laboratory Skills	2 years	/Lambton College	3.0	Course No. 1 or 3 and Course No. 2 + Lambton College
Advanced Water/ Wastewater Treatment	3 years	Lambton College	3.0	Course No. 2 or Course No. 5 + Lambton College

Phase III

- advanced laboratory course
- attendance of course: "Application of On-Line Respirometry and Other Analytical Instrumentation to Process Control"
- wastewater course at community college (60 hr)
    1 year to 18 months

Phase IV

- eight Ministry Courses, duration one week each
    2 to 3 years

CONCLUSION

Effective training is not an accident—it is a systematic approach. The mechanics of planning the training itself leads to both a pre-startup program and continuing training for company personnel. We feel that proper training is the single most important tool in a well operated unit. Constructing, expanding and upgrading wastewater treatment facilities present numerous employee-related problems as well as technical difficulties. Inevitably, problems will occur during the initial startup period, and prior planning will pay off. To ensure a complete, cost-effective and productive operation, the following three factors should be considered:

1. **proper design:** to allow for process flexibility and controllability
2. **instrumentation:** (on-line and laboratory) to allow for process information to be readily available
3. **solid and established training program:** so that the operator can learn to properly interpret process information and apply it to a flexibly designed plant

For furthur information on this topic, see the Questions and Answers section on page 222.

CHAPTER 11

STARTUP CRITIQUE OF SARTELL 3

David L. Keller

   St. Regis Paper Company
   Sartell, Minnesota

INTRODUCTION

The expansion was set up for completion over a 2-year period. The most important aspects of an expansion are the design and construction. If these two things are done well, startup will go well.

The expansion of the St. Regis Sartell mill consisted of a new wood room, TMP, coating prep, kraft pulping, a 420 ton/day paper machine, kraft and paper warehouse, 1450 psi power boiler and a treatment plant comprising a $350 million expansion.

The new treatment plant consists of one new 120-ft-diameter primary clarifier, two 1.65-million-gal aeration basin, two 130-ft secondary clarifiers, a 65-ft gravity thickener with a 24-ft × 24-ft reaeration basin.

The first flow prints of a new treatment plant are of great importance. With the first issuance of prints, the majority of changes are made to the process design.

It is a good idea to visit other plants that have comparably sized treatment plants. By doing so, you can get ideas on what is good and what is not, as far as a plant design and the problems to look for during design and construction of your plant. Problem areas are: equipment and construction design, instrumentation and process flow design. These are helpful to know when purchasing equipment for the construction of the plant.

It is also advisable to get to know your engineering firm and the people in charge of the design of your plant. You will be working hand in hand with them through changes and problems.

The following are just a few of the changes that were made from the original prints (see Figures 1 and 2):

1. process changes to the secondary splitter box by adding a splitter gate to have the capability of running two completely split secondary systems
2. changing of addition points of nutrient feed systems
3. putting drain lines on the aeration basins
4. addition of a gravity thickener and reaeration basin
5. changing pump design on chemical addition systems
6. changing piping design on return sludge pumps
7. sample points within the system
8. back flushing and flushing lines on all pumps and lines

VENDOR MEETINGS

Be careful about purchasing any equipment that has been on the market for less than one year or that doesn't have some type of a track record. Also, be careful when purchasing meters and instrumentation. Sometimes vendors are giving good deals on closeout items, and you may be stuck with something you can't get spare parts for. At the time of purchase of equipment from vendors, get spare parts lists (or purchase certain spare parts) and installation prints.

CONSTRUCTION

The construction of the treatment plant was one of the first projects started within the expansion. It was built in two stages, beginning in March of 1980. In the first year (stage one) of construction the #2 (north) aeration basin, # 1 (east) secondary clarifier and the secondary pumphouse were completed for operation. On May 12, 1982, these projects were put into service.

In the second year (stage two) the # 1 (south) aeration basin, #2 (west) secondary clarifier and # 3 primary clarifier were done by fall of 1982.

For winter the above units were put into a winterizing program in which secondary effluent was recycled from the operation side through the nonoperating side to keep the units from breaking due to frost heaving. This was done by temporary piping from #1 secondary clarifier to the boiler wet well, and the primary effluent line from #3 primary clarifier was taped into the bottom of #1 aeration basin.

The boiler wet well lift pumps were used to pump secondary effluent to

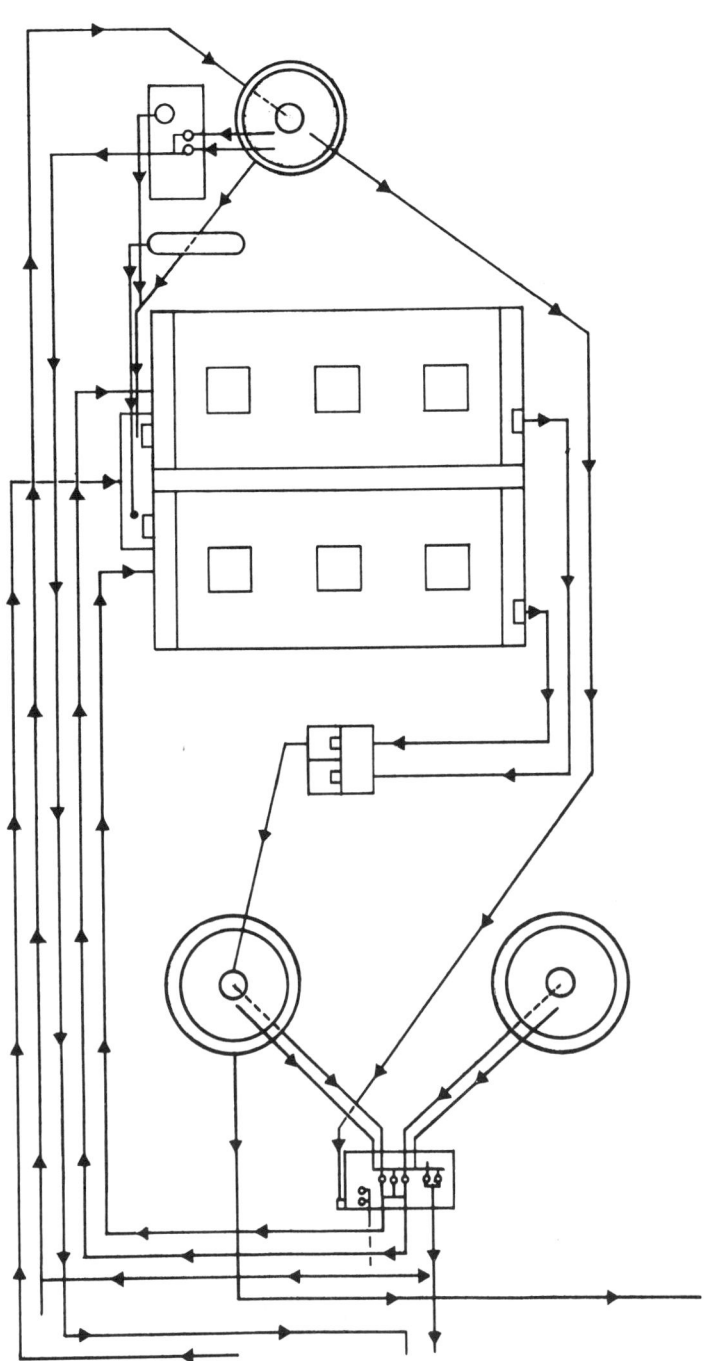

Figure 1. Initial treatment plant design.

176 ACTIVATED SLUDGE PROCESS CONTROL

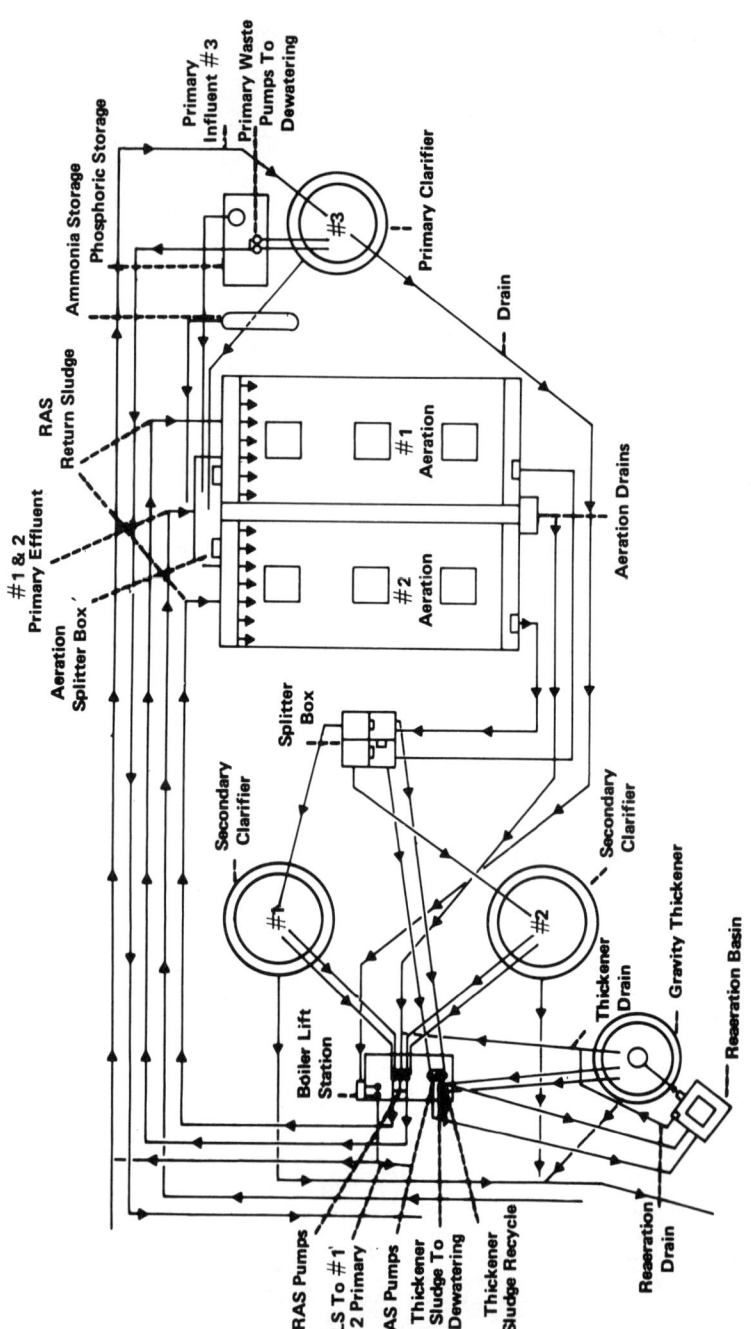

Figure 2. Current treatment plant design.

#3 primary clarifier and the flow from #3 primary gravity flowed through #1 aeration basin and #2 secondary clarifier.

In April the second-stage units were drained to repair leaking pressure ports in the floors, and the temporary piping from the winterizing program was capped off.

In April 1982, the contractor began work on the reaeration basin and gravity thickener, and completed the work in June. The operation was put into service on June 24th.

STARTUP OF STAGE I

Startup of stage I (east secondary clarifier and north aeration basin) took place on May 12, 1981. This startup was the most critical due to the transfer of the biomass from the existing secondary treatment system to the new secondary system.

Prior to putting these units into service, they were filled with secondary effluent so that the dissolved oxygen (DO) and nutrient levels could be raised to accommodate the biomass from the existing treatment plant. By doing this prestartup fill of the aeration basin and secondary clarifier it allowed us to flush lines and check out nutrient feed systems, pumps and aerator drives to assure that everything was up to par.

The transfer took approximately 4½ hours using two 2000-gpm (diesel) pumps. The biomass was pumped from the existing aeration basins to the primary effluent launder box of #2 primary clarifier in which the biomass flowed by gravity to the new aeration basin. The existing secondary clarifiers were emptied into the aeration basins by the recycle pumps.

STAGE II—STARTUP

The #3 primary clarifier was put into service when plant flow put too much demand on the two existing primary clarifiers i.e., at the beginning of departmental startups within the expanded mill.

The secondary startup (of the south aeration basin, and the west secondary clarifier) was done by the building up of biomass in the operating side of the secondary system to the point at which DO was marginal (0.5 ppm). At that point the nonoperating basin (#1) was started to be filled with primary effluent and the recycle sludge from the operating secondary clarifier (#1) was diverted to the #1 aeration basin. During the period of time in which the basin was filling, the mixed liquor suspended solids (MLSS) was monitored so that when the MLSS was reduced by approximately half, the recycle sludge was returned to the operating basin. Recycling was started at that point from the #2 secondary clarifier back to the #1 aeration basin.

SAFETY

Safety during an expansion of this magnitude is of great importance. During the expansion all operators were required to wear hard hats at all times when on the job site. There was a common lockout procedure on all equipment between construction and mill personnel.

MANPOWER AND TRAINING

Manpower during the project was a major problem due to the size of the project and union seniority. Prior to the project, the treatment plant was under the power plant supervision. At the beginning of the project it was made into a department of its own. The permanent operators moved into the line of progression in the powerhouse. This opened up positions for all new operators in the treatment plant. Due to union seniority, there was no hiring of operators from outside.

Manpower requirements should be studied very closely to find out how many operators are going to be needed for startup. In the case of doubling the size of the plant your manpower requriements may double. This may be only through a startup period, but it helps if you have the extra operators on hand. In the case of this expansion there were only four new operators and no spares going into the start. By the end of the startup the manpower requirements had doubled and training of operators had to be done in the midst of the startup. (Note: Startup is a good time for operator's to get experience; but it is no time for training operators.) This caused a time problem with getting new operators trained and getting training manuals completed for the new treatment plant.

Training manuals are most helpful in training many operators over a short period of time, but there is also a disadvantage of trying to write training manuals too early in the expansion, due to design changes in equipment, piping changes, or lack of prints of areas not completed. Training manuals should be set up in a catagory order that describes the plant from the start of process flow to the end. They should describe purpose, description, startup procedures, normal operating procedures and shutdown procedures. Safety should be included in the writing of the operating procedures in such a way that it stands out. Print it in red or put a box around it. Troubleshooting should be in with each operating procedure section.

PROCESS CONTROL

The process control of the plant is fully automated. The complete system runs off the Foxboro one video-spec analog computer.

All process changes are done from the video-spec CRT in the powerhouse and treatment plant control room.

Return sludge rates are set up on a ratio of 55% of flow through the plant, giving a retention time of about 1½ hr in the secondary clarifiers. This flow is controlled by the computer by monitoring the flow to the river.

Mean cell residence time (MCRT) is maintained in the secondary system by continuous wasting of mixed liquor coming from the aeration basins to a reaeration basin and gravity thickener. This is calculated by the following equation:

$$\text{MCRT} = \frac{\text{quantity of microorganisms}}{\text{rate of loss of microorganisms}}$$

In case of the shock loads to the secondary system, be it organic or toxic, the secondary system has the capability of being reseeded from the gravity thickener. The retention time in the gravity thickener is maintained at between 3 and 5 tenths of a day to keep the biomass most viable for possible use of reseeding. Recycling from the thickener to the reaeration basin is controlled by consistency of the thickener sludge trying to maintain 4% solids. Recycling is never stopped at any time even if the consistency drops below the target, due to the possible septicity and odor. The wastage off the primaries and the thickener is pumped to two belt presses for dewatering and incineration in the boiler.

From the first of the year through startup, (June, July and August) of the paper mill, the treatment plant efficiency of biochemical oxygen demand (BOD) removal never dropped below 95%, and averaged 96% from January to September (Table I).

PROBLEMS DURING STARTUP

The problems that were encountered during startup were major. The first problem was discovered before the mill startup. It was found after draining the #2 (north) aeration basin that all the lower pumping turbines

Table I. Treatment Plant Efficiency

	Primary Effluent BOD #	Secondary Effluent BOD #	% Removal
January	170,097	7,616	95
February	98,809	4,641	95
March	137,396	4,556	97
April	150,699	6,780	96
May	219,954	6,016	97
June	372,943	18,296	95
July	479,980	22,289	95
August	435,798	19,422	96
September	327,360	14,183	96

had fallen away from the upper turbines. The shafts connecting the upper and lower pumping turbines were bent and had to be straightened, reinstalled and torqued. It is still a mystery as to what caused it.

The startup of #3 primary clarifier was the second problem encountered during the mill startup. The clarifier drive tripped out after 3 weeks of operation on the line with 4 ft of sludge in it. It was found to have badly worn pinion ring gear, but this was not the whole problem. After the gear was changed, the clarifier rake drive went down two other times with 2 and 3 ft of sludge in it. The problem was solved by cutting of the rake arm scrapers so that there was a 3-in. clearance under them. Before that the scrapers rode on the surface of the floor.

The third problem encountered was thickener sludge pumps. The pump would not pump 4% thickened secondary sludge to the dewatering blend tank. The pumps were positive displacement moynos. The solution to the problem was the installation of larger starters and barrels and larger motors.

The fourth and final problem was with the sludge dewatering presses. The presses had a lot of roll failures in the pressure sections and belt mistracking on the gravity sections. The presses were down quite frequently due to these problems. The manufacturer of the presses was brought in and the presses were actually rebuilt on the run, so as to try to keep up with the sludge inventories. During press down time, the existing coil filter dewatering system kept the inventories out of the danger zone.

SUMMARY

Inspite of the problems that were encountered, the startup went very well. The plant had no monthly violation and was meeting its 1987 permit conditions.

CHAPTER 12

ROUND TABLE DISCUSSION

All Attendees

Vernon Stack: The discussion this morning about BOD respiration—I think that's a good one. I think it's a kind of thing that we're all interested in. It's not a new subject—it's a fairly old subject in the sense that some tests on respiration of some kind have been made for various purposes for a long period of time. And the respirometer's been around for a long period of time. And there were people building respirometers eons ago. I made one point after the discussion this morning, but the thing that we're looking at when you're talking about respiration is that period of time when the organisms with the new food supply synthesize that into new cell material. As soon as they get through that synthesis we're back to the endogenous level. Endogenous level might be a little bit higher because we have more cells probably than we started with if the measure of food was significant. In the curves that Dr. Therien presented this morning, one thing that I saw, and I've always looked at Bob Arthur's information, I've always sort of run this through my mind, but when you're taking a given load of food to organism in that respiration and you're cutting the respiration off at one particular point. You're cutting off a piece of the total energy oxygen curve. If you cut in 30 minutes, depending on how you put it together, you can put it together so that the whole energy oxygen is complete in 30 minutes and if you do so, then you've got a linear relationship. If you put it together so that the respiration of energy oxygen is going to be 3 or 4 hours and you cut it in 30 minutes you only have a small piece of it. The rate would be high but it's there. So that in going through that when you select a time to cut it you're taking a piece and if you select some specified time when you may not—unless you elect to put it together so it's going to be completed maybe in a certain period of time by dilution or whatever technique—you're throwing in a variable, you're slicing a certain piece off. Then you get the curve. So I think just to understand that the energy oxygen that's there, that measurement is very precise, in fact it's the most precise thing I know of in bug behavior in that it will take exactly the same amount of oxygen to stabilize exactly the same amount of material. No matter how you put it together that will happen. I just wanted to comment and get that started and see if anybody had—I'm sure that the two doctors will have some more to add to that. Anybody else who had some experience that fits in, I think it's an interesting thing to dwell upon.

Dr. Robert M. Arthur: Dr. Therien, in regard to what Smokey said, the studies that you seem to be showing the most of had solids concentrations of around 6000 as I remember. Now Smokey is correct that if you cut it off in a short period of time, you're not going to get as much oxygen utilized in that short period of time, and therefore correlation with BOD isn't going to be as great. When we first started out 15 years ago, we cut them off at 1 hour, well we actually ran a full curve, but then we analyzed only the first hour, the second hour, the third hour, and the fourth hour and we got different results, we got *less* correlation the shorter the time. Now this was with unseeded samples, though, and you're seeding them with 6000 milligrams per liter of solids which we don't really know how much viable organisms are in there, but certainly there's more than there is if it's not seeded. So that you should be able to shorten the length of time the more bugs you have in there. What do you feel is an optimum value to use as far as suspended solids in this kind of correlation test?

Dr. Normand Therien: Well, this is highly variable on the strength of your liquor. Obviously some of the tests that we've done with beef extract or industrial waste liquor you could repeat the test for example working with 3000 milligrams per liter than work with 6000 milligrams per liter VSS, and you'll find out that it will take less time. But it's hard to say really what you would use. I would recommend to use typically sludge that you'll find in the aeration basin. If you're operating normally, you'd probably have around 3000, 5000 or 6000 and work with that. What you'll find out is that you'll have to do some experimentation with the substrate; you have to find how fast it responds and then to go from there and say, "Okay, we'll establish a kind of calibration curve." 'Cause I didn't mention that much this morning, but really on a day to day basis over weeks or seasons you find out that your distribution of species, your solids varies and obviously it will play a role in the respiration rate, it doesn't exactly respire the same way. So that you'll have to do some experimentation in terms of working with the given solid concentration. Then taking your waste—to dilution for example—do multiple tests and get a kind of calibration curve. One thing I didn't emphasize this morning but I would like to do right now is that through microcomputers if you want automation you could update your calibration curve over time just through routine plant BOD_5 analysis each time you get a new—you could take for example a sample which is fed to your respirometer and then do a BOD_5 analogy on it and each time you get a new point you kind of follow that in time. So that's to update your calibration curve and maybe leave off some of the date point's which are three weeks old or something like that, and that permits you to really have a little bit of reliable kind of calibration curve to predict the BOD_5 if it is important for you to predict BOD_5. Because you could really use directly that measurement coming from your respirometer.

Dr. Robert M. Arthur: I think that's a very important point and one that we've been trying to say is that—well, you could say the same thing about a settling curve—why try to generate flux data and all this kind of stuff when the settling curve tells it as it is, and the respiration curve tells it as it is and that in itself is process control information. When people say, "Well, how does it correlate with 5 day BOD?" I say, "Why do you want to correlate with 5 day BOD?" As you pointed out, the accuracy of plus or minus 20% on a BOD analysis is going to throw your curve off right away. And if you believe what Smokey said that energy oxygen, which is the amount of oxygen used to metabolize a food, is a constant—in fact, if you got a constant food you can calibrate the instrument almost if you can stoichiometrically determine the amount of oxygen that's required. You can almost calibrate the instrument on that value.

Vernon Stack: In following your precedure, it seemed like you would not like to get out the concentrations where you were tending to get into the curve. That's where you'd like to use dilutions, as you suggest. Would you have maybe, if you have quite variable strengths in the incoming wastewater, would you need some other parameters such as a quick TOC or something so that you get down to practical dilutions? Or is that complicating it too much?

Dr. Normand Therien: We had the problem just working at the plant when we said, "Okay, we'll go for a 24-hour campaign." At some stage you could really, if you want to go automatic about it, you could devise a scheme based on your respirometer. You're finding out that the oxygen uptake rate measurement is too high. This kind of invalidates that test and for the following test go to a dilution. This can be done automatically. It becomes a little bit [French word]—but I guess you have an equivalent I'm sure in your language. At the end you have a lot of preconditioning and a lot of presampling conditions to satisfy, like a pH adjustment, like temperature adjustment. In the end it becomes a lot of things to do. But we're using on-line TOC and surely we could easily devise a scheme where you could use that information even if what's coming out in terms of TOC is not maybe exactly in the same ratio in terms of biodegradable material. But surely it could be devised. The very nice thing that I get amazed at with the respirometer is the fact that it doesn't give you an absolute number answer when we're talking about BOD_5 relationship. But it is something that evolves with your process. If you have different types of waste, different fraction of biodegradable material, there shall be evolution in your species because if you're operating at $5°C$ in the wintertime and $20°C$ in the summertime, it may affect your distribution of species and you just don't worry. You just pick up the measurement from your respirometer which operates with the bugs that are in your plant, operates on the type of liquor that you want to treat, and most important, usually you work with solid to liquor ratio that will be near the one that will be in your aeration basin. The only thing is that maybe if you want to relate it to some more conventional classical parameter like BOD_5 then you have to adjust to the idea that you would have to have a calibration curve which will vary in time. Maybe I'll comment later on that in terms of temperature because this is something that plays quite a role in the respiration rate.

Walter Frais: Basically what you're saying then is depending on X number of variables BOD, respiration rate curve will change almost continuously. And you can never establish a standard curve which you can say six months down the road will apply. We've gone through the . . . we tried to correlate COD and TOC and we tried it for nine months and we got so many variable chemicals, it's just we couldn't do it. You get a nice number which might change from day to day, but I can't see establishing a curve in your case and staying with that curve forever. It's almost an ongoing thing.

Dr. Normand Therien: We're presently working on that at the plant. What we're doing is each month we're establishing a calibration curve and following that. When establishing that calibration curve we're fixing some of the parameters. For example, we're doing it always at the same temperature because what we thought of one time is, "Okay, we'll store some of calibration curve, obtain a different temperature, and work from there." But it doesn't work. In time you've got so many other parameters that vary—so we're fixing temperature, also always adjusting to the same pH. What we're doing is followed over a whole year period. 'Cause we're—I feel strongly on that—in the academic world we have some facility, we have students and we can maybe spend some time on the basic things. At the same time, sometimes it doesn't take that much of our own time to go and try to answer questions that will be pertinent to the industrial world. For example, what about calibration curve. How does it go in time? How is it reliable in time? Now at least for next year we'll have a set of 12 calibration curves that will at least follow and we'll tell you at that time if we really got something out of it or not. What we're trying to do is simply say one thing that is changing is probably the substrate distribution in time and the bugs also. So okay, since the respirometer takes these two parameters into account let us see what type of result it gave us as a change over time, maintaining other parameters constant— the one at least that we see.

Walter Frais: What will happen if you take bugs which are in the middle of the winter and they are at 10°C—your system's running at 10°C—and then you'd go take them to the lab and you run your tests at 20°C. I mean you've got the wrong species. How can you relate that respiration rate to a system which is at 20°C?

Dr. Normand Therien: These are typical of the types of problems I have to discuss with some of my students. To give you an idea, they will have the tendency to take these bugs and go and raise then to 20°C in 15 minutes, even some people just putting a pot on a stove just to bring it to 20°, you don't do that. This is one type of problem that we're facing right now is how would we go in terms of procedures? What we're doing right now is saying, "Okay, we'll give them 24 hours acclimation. I'll want them to come to the endogenous phase so I just leave them wholly there. We'll probably do the same thing when we go up to 3 days and 4 hours acclimation because, as you pointed out, this is important information. How do we go with that? How does it affect? The other thing that you have to go through if you want a calibration curve is you want to really do a parameter like BOD_5. Otherwise you could use the data first for itself to control. My thought this morning was simply relating to BOD_5 because I found out that even when I talk about carbon, people are relating to parameters that they know and even if some of you may treat BOD as a mysterious variable, it is well used and sometimes you have to relate your own result in terms of that variable. If you recall Dr. Schroeder yesterday arguing about the VSS, for example, and BOD_5 always relates to respiration data. Well it does relate—we could show that on the resurgence of condition. It does relate. So it permits the discussion to show that since it relates you can maybe transpose some of the results in one field and to the other field. These have some correspondence. It's not equivalent.

Terry Charles: This is directed to anyone up there as long as we're on respiration. We talked about temperature control in the testing, the type of food substrate we give it, and using a seed. Are there any precautions or warnings that a person should look at in the seed itself, as that changes from week to week? If you are going to use an on-line instrument for process control do you run some calibration checks on the seed or is there some type of procedure to check the condition of your seed? Is it anoxic, taking it off the return rate?

Dr. Normand Therien: I'm sorry. Could you repeat some of your question?

Terry Charles: I'm concerned about the seed itself. What type of condition it's in.

Dr. Robert M. Arthur: In other words, going from week to week, Dr. Therien, or day to day, are the return sludge characteristics, or whatever you're using for seed, going to change significantly, and will that affect correlation?

Dr. Normand Therien: My feeling is that it will change in time. At least I'm making an assumption that it will change in time. So we're making an assumption that it will change in time and we're kind of just checking how you would go along and if we can establish a reliable procedure that will permit us to come up with a kind of calibration curve that could be reliable. Surely your distribution will change, like we talk about the temperature affect, if you're doing a respiration test you'll find out that even in a lab under very controlled conditions a 1- or 2-degree variation will affect, will be quite a change. So we're just kind of . . . we're making the assumption that it will change. From our experience we know that the characteristics of a population of bugs will change over time. Species or measurements at different temperature, for example, even acclimating to a new temperature. What we're kind of looking at when we're saying we want to go and check a different temperature we are kind of making the assumption that the distribution of bugs and species won't change that much during the period you are raising the temperature—that is about over a day period. Otherwise if the distribution of species starts changing drastically then you would be measuring something, but you have a calibration period that will be different than what would be actually occurring in the process. The bugs will come from the process so they'll have essentially the same distribution in terms of species and you can make tests over a short period of time at a given temperature to see how this affected the test. But the test will always be done at the same temperature.

Vernon Stack: If I'm reading your question correctly, I think you're concerned that maybe something has happened to your biomass on a short-term basis which makes it not as viable as it was before. You might consider, as Dr. Therien seems to think appropriate, you might consider having available a source of food material which is not the wastewater but is something else which you've prepared which is compatibly like your wastewater, hopefully, that you can give a load of that to the microorganisms and see if you get the response you expect. And if your organisms are about the right population and the number of viable organisms are there and you give it this particular fixed dose you are to come up to an uptake rate that would be a certain value. You could be more complicated and maybe try to make an ATP measurement a load or something like that, but I think that's probably getting fairly fancy.

Dr. Robert M. Arthur: What's normally recommended is that if it's an on-line instrument then periodically you introduce a sample of tap water instead of your sample. And then you're monitoring just the seed and (depending on how frequent your seed may be changing) you can do it once every 4 hours or maybe once a day but if you see a good consistency in respiratory activity with just the tap water or maybe your final effluent is a better thing to add depending upon how bad your final effluent is, but that would give you a norm, we call that the baseline, and then if that changes, you could pick that up. Does that answer your question, Terry?

Terry Charles: Well, what I wanted to analyze is the final effluent, trying to find some type of correlation with BOD_5 going out.

Dr. Robert M. Arthur: Well, then you'd just have to use just tap water so there wouldn't be any uptake at all. But we've found that you can, like at Wisconsin Tissue Mills, they're just putting their final effluent in with the seed and you can see that stuff go up and down with the BOD changes in the plant. So periodically they just do it manually, they just introduce just the seed itself. You can do that too. You don't even have to mix it with tap water, but you just introduce the seed and if that changes, then you know you've got a change in the seed, providing that it's endogenous. It has to be endogenous.

Dr. Normand Therien: Maybe just to add something here, we talked about kind of going at reference or standard type of condition, to see the evolution in time of, you know, your bugs in terms of respiration. And you use tap water to go with the lower line, but as we discussed this morning, for example, you could use as a biomass-related parameter RP. It probably doesn't mean anything right now than to me but the respiration rate that you would measure when you are submitting your bugs in the endogenous stage to a reference, and this is something that you can prepare and be very specific on, that in terms of procedure using beef extract or glucose or anything, and this is what we've done in the test repeatedly is do a simple—not half an hour but ten minutes—measurement using a given quantity of beef extract and when I'm saying a given quantity, is a quantity that you make sure is an overdose of nutrients then they can consume. And you just do measure that. This is one reference that will follow in time. Even when we play in bringing your effluent which may be at 5°C and you want to do the test each month at 20°, even if you're able to raise the temperature in a few hours or a day we'll always use that standard just to see how RP evolves in time. And then do our test but that's a great idea. You can go with a low reference using just plain water or anything, but also you can go with checking the active part of your biomass, and this is what you want to get at, the active ones.

Terry Charles: I was thinking that it would'nt be necessary then to analyze the seed side by side as you run respiration rate tests with the effluent. I'm looking for a rare thing trying to predict outfall BOD_5, and I want to be as accurate as I can. But it wouldn't be necessary on every test, whether it be every 30 minutes or every 4 hours, to analyze that seed. That's a constant parameter.

Dr. Normand Therien: As Mr. Stack was mentioning, here is essentially once a day. And this is something that really you would feel for yourself. You'll probably experiment, "Gee, I don't have to do that test that often because it doesn't change," or you may find times in the year because of bug succession or anything that it becomes critical. There's a definite change each day or so or three days, there's a trend, and then you do a little bit more. But the nice thing about it is that there's no basic rules—you will have to feel it a little bit. Our experience is that you won't have to do it that often.

Don Krohn: I'm curious. I'd like to hear some discussion on the future of using respirometry as it relates to process control in municipal plants. Anyway we don't have a lot of control over our influent BOD. It's nice to know what it is coming in but do you folks see any future in tying in our respirometer reading to a variable-speed return pump and possibly maintaining a constant oxygen uptake rate that once you establish an oxygen uptake rate that's good for a particular time of year and a particular set of conditions, do you see any future in trying to control that uptake rate at a steady state for optimum treatment?

Walter Frais: Actually, I've mentioned before we're dealing with WTC— the Water Technology Center in Burlington, Ontario, Canada—and they've got a system on-line now that's got a computer on it but it's got a system on-line where they measure the respiration rate and measure other variables. They've got a little module which predicts KLA and they automatically adjust the amount of air they're adding to a system as a result of that. I've seen some of their graphs and they've got very smooth control. And the energy savings from such a system are quite tremendous. So if you're interested in finding out more of that, contact them.

Dave Keller: I was going . . . my question is, has anybody done any study as far as nutrient addition respiration rate, and using it as far as optimizing nutrient addition to an activated sludge plant?

Dr. Robert M. Arthur: The instruments have been used so many different ways, and once you get an instrument, you find new ways of using it. But I don't know specifically that anybody has used it to optimize nutrient dosage either on a lab or an on-line basis.

Vernon Stack: I was going to say that there's a basic problem in trying to optimize nutrients based on something such as respiration in that the problem existing and the respiration rate changing are not sequenced in time. I mean you can take all the phosphorus away and the thing may tend to appear very good for maybe even weeks after that and then it suddenly collapses. I guess the phosphorus is there no longer. It's maybe a little more closely related to the ammonia or nitrogen available. I've never really looked at it and never really have implemented it to any degree, but say just meter or pace the nitrogen feed to the uptake rate 'cause it's representing the amount of synthesis that's there and you've got to have that ammonia available right at that time. I think that's feasible to do, but if you cut the ammonia off, you probably won't see the uptake rate change for some period of time. It's just the natural sludge age and what's there and the flywheel is going on, it won't stop turning.

Robert Davis: Thanks, Smokey, you brought up the question that I had as I was very interested in. Have there ever been any correlations or any analysis on how the respirometer might be used in nitrification systems by perhaps monitoring the effluent by an on-line respiration meter in order to make some process control decisions, especially considering that the percentage of nitrifiers in the mixed liquor is so low? Any comments?

Vernon Stack: I was trying to see if I understood your question. As far as respiration and as far as nitrification are concerned, nitrification is proceeding very much in tune with the ammonia that's to be oxidized and so forth, so that I think respiration as related to the amount of, if you're using it upstream or something to try to measure the amount of load that's coming in and the amount of ammonia that's there to be oxidized, I think it's as good a measurement for that purpose as for carbon, but is that your question?

Dr. Robert M. Arthur: Bob, are you interested in monitoring the stage of nitrification?

Robert Davis: Well, we're installing an on-line ammonia analysis on our influent, and how that's going to be able to give us process control right now we don't even know. Right now we're just looking for the information. I'm wondering if monitoring the effluent from a respirometer for ammonia nitrogen based on the mixed liquor concentration that you have in your instruments might give you some basis for a decision in process control in order to change that effluent in the real system.

Dr. Robert M. Arthur: The instruments will not differentiate whether it's carbonaceous demand or nitrogenous demand. If it's oxygen demand, you're going to see it. I think a better way of following nitrification-denitrification is with ORP. Steve has done some real good tests following the—we did it both ways-nitrification and denitrification and you can see that ORP change depending upon—I'm starting to sound like Dr. Burbank from last year—but you can see that ORP change depending upon the stage of oxidation. That's all you're doing with nitrification. You're trying to oxidize the ammonia and you can see the stages that you're going through with different levels of ORP. And in the opposite direction—denitrification—you can see the same thing. So I think I'd suggest some experimental ... we don't—I'm not trying to promote

ORP—but if it's a tool, it's another biological measurement in addition to respiration, and I think it's a much ignored one and you can get an ORP probe for what?—80 bucks or something like that, Steve?—and stick it onto your pH meter and do a little experimental test with your nitrifying bugs and I think you'll see some interesting results.

Robert Davis: Will that give you the actual stage of nitrification you're in? I mean we're nitrifying all the time; we never have a problem of not nitrifying. It's just that because of influent changes we will nitrify it completely enough to meet the water quality standards and I'm not sure what process control techniques we're going to be able to use because you can only return so much sludge.

Dr. Robert M. Arthur: At this stage I don't know what it would show you. All I'm saying is it might be another tool for you to look at and try to analyze what stage it might be in.

Don Krohn: In Fort Collins we've been able to tell somewhat our extent of nitrification by watching the oxygen uptake rate, and it's a broad generality to make, but providing that most of the plant was in steady state you can actually see the oxygen uptake rate lower when you're nitrifying and higher when you're not. And it will lower generally to the degree to which you're nitrifying.

Dr. Robert M. Arthur: Very interesting. I'd like to ask Dave Keller a question. How did you feel when you were listening to the presentation by the two guys from Canada?

Dave Keller: Well I . . . like you'd mentioned before, I felt very envious because I had the biggest problem of trying to get operators to be present for startup with some kind of wastewater background. And I did have them for a short period of time, then I lost them and got a totally new crew that had never ever been associated with wastewater and I had to train them and I had only four months to do it in and during startup it created problems and that's why I spent so much time in startup doing operator work because of the fact that I had inexperienced people. And I think it's great what Polysar is doing as far as training and getting prepared for a startup because it'll make a much easier job for everybody all the way through.

Dr. Robert M. Arthur: I'd like to ask to see a show of hands of people who have gone through startup operation in a wastewater treatment plant. Okay, next question is, is there any such thing as process control during startup? Who wants to attack that? Any of you guys that had your hands up.

Dave Keller: Maybe I can add a little something to process control during startup. We had process control—it was due to automation. And as far as return sludge and nutrient addition and that, we kept monitoring our nutrient addition to make sure that we had proper amounts there. If anything else, we overfed them to keep them viable, and as far as return sludge rates, we put it in on a ratio block on our computer at 55% return rate of incoming flow and we let the computer take care of our recycling for us so that it was something we didn't have to worry about.

Robert Davis: Our process control during startup consisted of sand filtration. You know, it catches anything that comes out. But I'll say that when we started up our tertiary nitrification that nothing went as planned in the process. In fact, it started so easily it was unbelievable. It wasn't even two weeks when we were fully nitrifying. No seed, it just took off. It is very hard to theorize what's going to happen.

Don Krohn: We were both involved in startup at plants other than Fort Collins. We haven't been involved in startup of a plant in Fort Collins. We were both involved in startup of plants at other municipalities. It's always rough. The best you can do is plan, get some guesstimates on the characteristics of your waste and use your experience to match your mixed liquor concentrations and things like that and fly by the seat of your pants and hope everything runs.

Dr. Normand Therien: Yesterday Dr. Arthur posed the problem in terms of process control, you've shown that the possiblity of using a microorganism storage tank where you could store sludge there and probably recycle it at the proper interval to your aeration basin to do a specific job, for example if you had a slug of food coming into your aeration basin. We had some experience in the past year in deriving optimal control policy from data and I'm going to probably make some of the people afraid here when we're talking optimal control policy in terms of tools like maximum principal and things like that because we're saying, "Well gee whiz, we're not at that stage yet, you know, just leave us alone." But we found some of the policies which are very attractive in the industrial sense and what I would have liked to pose as a question here to the audience is: is anyone contemplating or has anyone used an ordinary storage tank for sludge in view of recycling?

Dave Keller: We use our gravity thickener reaeration for activated sludge storage. We use it for an F/M control, and by doing so we do have the ability to return our stored sludge back to the aeration for toxic loads or organic shock loads and it works very well. For startup it was I would say the major thing that saved us many many times due to the fact that with the thickener system—thickener reaeration system—we could waste mixed liquor on a constant MCRT and it was another controlling factor that we didn't have to worry about. We set up with a MCRT of 4 days, we kept it a little high in the summer months due to the heat and the oxygen demand, we kept our sludge a little older, we like to keep it around 3, 3½ days, that's where we get our most optimal viable range. It was another helpful process design that really got us through startup due to the problem we had with our presses and getting rid of sludge. It was a place to store things. so it works very well.

Walter Frais: Pursuing that a little further, how long can I realistically store sludge and then return it to my system and will it be effective? And besides air, what else would I have to add?

Dave Keller: Well, realistically we don't like to store it for more than—keep sludge for more than 2 days in a system. We don't have any type of a nutrient feed system to our thickener-reaeration. The optimum storage time on a thickener reaeration I feel is between 3 and 5 tenths of a day. You like to keep it turned over, keep it most viable for reseeding and we have approximately a 2- to 3-day storage period. And within that period of time you can get yourself back on your feet in most cases.

Dr. Robert M. Arthur: We've designed I think, John, two plants or three plants with microorganism storage? And we gave a paper on this up at the WPCF meeting. But as far as how long can you keep the sludge in there, Steve has taken sludge from an aerobic digester and that sludge is still viable if you feed it. And actually this microorganism storage could serve as an aerobic digester 'cause you're adding air to it. Now I would suggest that it be fed periodically and I don't know what periodically means but with respiration tests you could continually measure the viability of the system and tell from that how well they were being fed, if they were dying off. Food to microorganism is one use of microorganism storage. But in Wisconsin most small plants are wiped out in the spring because of infiltration/inflow. We have a lot of leaky sewer systems and we had a lot of unscrupulous contractors 50 years ago who put the pipe in the ground but didn't care if it had leaks in it or not. But in the spring you get a complete washout of all the microorganisms and it takes time to regenerate. During a 3-month period they don't meet standards because you can't ... once you get the bugs built up—another washout. Build them up again—another washout. So we said, "Why don't we during a rainstorm store the bugs. Immediately after it stops—bring the bugs back into the aeration tank." That takes a little kind of piping arrangement to do all that but it's possible. Of course we fixed some of the sewer lines that are going into these plants now so I don't think we've used that system once yet, but it was a way to eliminate the problems that are related to infiltration/inflow.

Dr. Normand Therien: I have a question to Dave. When you're recycling sludge like that, I assume it may represent large quantities sometimes. What about your clarification unit dynamics at that time? Do you have any problem or are you finding that the solids that you're recycling at your aeration basin are causing you a dynamic problem in your clarification unit?

Dave Keller: As of yet I haven't found any problems with it. We've had the unit on since June 24 and we've stored up to 2½-3 days at the max and right now within the thickener unit or since that time our effluent coming out of it has just been crystal clear. There was one case in which we went in to denitrification due to the aerator failure, but fortunately we only had a minimal amount of sludge blanket in the thickener and we had the cake popping to the surface. And due to the fact that the aerator failed is what the major cause of the problem was. Other than that, that's the only problem we've ever had as far as quality coming off that unit.

Robert Davis: In our treatment plant we had no thickening, so when we were building our nitrification facilities we were worried about shock loading the treatment plant with supernatant from our digester. So we built extra aeration tanks to aerate supernatant and we found out that this will give us a culture of nitrifying bacteria that we can use during storm periods as a seed, and we're also investigating the possibilities of feeding it continually to raise the load to the nitrification facilities so that when there is a shock load we can pull it away and they will already be acclimatized to the higher influent.

Dr. Robert M. Arthur: Listening to Bob there, it makes this so interesting because every plant is different and you've come up with a little idea that is gonna perhaps make your plant work better, and he's got microorganism storage which not too many people have but every plant is different and the problems are different. Anything else on microorganism storage? Any questions on that? I want to change questioning and ask Bob Skrentner, we've talked a lot about the computers at this conference and I want to ask you the question, Bob, I was surpirsed at the number of hands that went up when you asked the question on how many had computers or were going to put computers in and how many hands there were up. Do you see a trend from the engineering profession because your company works primarily with consultants, right? Oh, you do. Okay, but do you see a trend in any kind of imagination or innovation in the engineering profession to use the computer in its full capabilities rather than just as a data logger?

Robert Skrentner: You're putting me on the spot. I don't want to say anything about my clients. The trend I see is more I think from the point of view of the owners. They get a computer and it can do a lot of things. And I think there are, let's see, Milwaukee left, but they've done a lot of things with their computer, Denver's done a lot of things with their computer, Detroit has done a lot of things with their computer. Little Empire, Minnesota, which is a 7-MGD plant, has done a lot of things with their computer. And you've done things with your computer and you're still working on it. And I think a plant that's managed well and has a dynamic leadership will use whatever tools they're given to the fullest and if the engineer or the vendor doesn't give them such a good system, they'll make it work anyway. I've seen a lot of Cadillac systems that aren't used and it depends much more on the plant, I think, than the engineer or the vendors. I personally don't believe the Milwaukee computer was state of the art when they got it. It certainly isn't now, and yet those people really use it. So that says something for the owner. So I think it's more dependent on the owner and the attitude of the owner and how they go about using the tools they're given. I really wouldn't say too much about the engineers.

Dr. Robert M. Arthur: That's a good answer. You didn't offend anybody there. Another question along that same line, "Do you see a lot of instrument people coming out with microprocessors on their instruments? Do you see that as a trend to have distributive type control VS the one central computer?"

Robert Skrentner: The trend is no doubt toward distributive control. I sometimes worry about also a trend toward microprocessor-based controllers right at devices. I worry about that to some extent, and that to me is almost getting back to analog controllers out at panels. You have a valve and feedback and you've got a little microprocessor—that's the same as an analog controller. And if somebody wants to call it digital that's fine. And I kind of think that's getting too far away from digital and plus, just for people who might be thinking of putting something like that in, there are interface problems with those things, especially if they start sending digital bits down highways. It's nice 'cause it locks you into a vendor, but I wouldn't, I'd be a little concerned about that. But I see the trend like toward the smaller distributive systems and for smaller plants I know some of the people were thinking of some smaller treatment plants, and I think when you tend to think of computers they think of gigantic $5 million things and whatever. You can buy a good basic unit for about the least expensive I've seen is about $12,000 and a faily decent one for about $30,000 and the nice thing about them if you buy them from the reputable vendors—and there's a number of those around—you can build. It's very common in industry for instance—I was at Fisher Controls. The industry starts with a $12,000 basic system. And the thing is fully expandable forever. The same with Foxboro, the same Leeds and Northrup, the same with almost any of them. So you get hands-on experience with this thing you put it on three loops or something, but you find out how it works and you do your own stuff. As you want to expand you go out and buy another one. And it's very nice for operating budgets. You can gear these things over a long period of time, and so for the municipal users I'm big on changing their attitudes. Don't acquire everything at once. Do it I think if you can, take a little more of a bit by bit approach and phase things in and extend it out and get the experience, and also with the operators I think I mentioned that, they have to learn how to use these systems. I think distributive systems are great because you can start cheap and build. And you don't need a big central computer to do all your logs and operating reports to start with if you don't want to, just get the basic control going. And you don't need optimizing which I talked about yesterday, that's like 5 years after you've started up. Maybe we like to set the trends a little bit, but . .

Dr. Robert M. Arthur: Are there any comments?

Pat Quinn: I just have one comment about building. It seems like that if you start with a little bit and keep building on it, if you're adding sensors and everything, you have to have some forethought so that you have conduits and everything (wires) going all over your plant, it's quite expensive. Just to start you might be better off to start with an overall concept of what you want to have so that you can do most of the construction as far as sensors and wires going everywhere at one time. Or did I miss the boat?

Robert Skrentner: There are different ways to design those computers. In most sewage treatment plants there are area centers, for instance secondary treatment you usually have a building or a gallery or something where a lot of your pipes and valves are and somewhere over by clarifiers you usually got something over there and it's usually a motor control center. And what you can do is start off with a small digital system in each of those areas and don't even tie them together at first. And if you decide you like it or have the budget for it you can eventually tie those together with a data highway. Now a data highway is normally—well, depending on the vender it can be a 6-inch conduit, but usually it's something smaller than that. So then you can run a wire through the conduit sometime in the future if you wish. In other words, you don't have to have a gigantic master plan and your $10 million all laid out for the next 50 years if you don't want to. I think you ought to think about that first, but you don't have to. You can buy, I think Milwaukee has just got one of the vendor's units at their plant, in fact two different types, and they're just hooking them up and playing with them. And they're very experimental that way. They like to play with the equipment, but you can start that way. A lot of people do without really a master plan. I would prefer that there would be one, but you don't need one.

Dr. Robert M. Arthur: Pat, you're from Metcalf & Eddy right? About 2 or 3 years ago I saw a demonstration by people from Metcalf & Eddy using a computer to demonstrate the activated sludge process or as a training device. Do you know what I'm talking about?

Pat Quinn: I don't think it was as a training device. I know we submitted a proposal to the state of Massachusetts to develop a computer training aid for I believe it was activated sludge. But I don't think there's much being done on that subject. We did have some demonstrations and we still do on our time-sharing system in Boston.

Dr. Robert M. Arthur: That's what I'm talking about. This was at the NETA meeting up in Minneapolis—the National Environmental Training Association—and maybe that's where I got confused on the training, but it is that, what did you just call it?

Pat Quinn: Well, we call it our RODA system, for Records and Operational Data Analysis, and it's very similar to the Envirotech system except for the fact that currently our system is on a time-share system and we go through a GTE telenet data transmission back to our Boston computer, and Envirotech is on Apple computer. And right now we're developing an IBM microcomputer, the same system, and we also offer a computerized parts and equipment inventory, management, preventive maintenance, scheduling system on our systems. It seems like I've been in different areas in the demonstration of these systems and it seems like there's more and more companies coming out with the same concept and it seems like it's a cheap way to go based on, you know, without getting into the full process control but just in the records keeping.

Dr. Robert M. Arthur: You're actually renting time on a computer, right?

Pat Quinn: Well, it's actually we lease our time. We sell the time on our computer, but in about the next 5 months we'll have a system where we'll market our software and you can buy a microprocessor, a 16-bit microprocessor.

Dr. Robert M. Arthur: Any other questions on that topic?

Robert Davis: We did it kind of opposite the way Bob Skrentner mentioned in that we got the data logging system and wired everything up to it and it sat there and collected information for the first couple of years before we even thought of installing any type of process control. You know, it does take an additional wiring but in some ways we prepared for that, hopefully, because we just need two wires in the building and we put a full cable of 22 wires and just let them hang there.

Robert Skrentner: You have a central computer with yours? So you need to hard wire everything back to it pretty much.

Lane Keffer: One of the reasons that I'm here today and yesterday is looking at the program I noticed some discussions on computerization at waste treatment plants. I was hoping that I could find out what kind of life that you see with these systems in the corrosive atmosphere that you have that we all have to live in. In an industry where we have a lot of acid fumes, we have a tremendous problem with instrumentation. But we recently have computerized our boilerhouse with DDC. For optimization we have everything on optimization—our boilers, our turbogenerators, our purchase power, low-temperature refrigeration for process, and it's done a superior job. A lot of our process is computerized already. Not to really give Fisher a plug, but they are Fisher controls, and we have full analog backup in our boilerhouse, but they're DDC. What kind of corrosive problems are you seeing at your wastewater treatment plants with computerization? And how are you protecting it?

Robert Skrentner: That's I think a question that every client probably asks us when we talk about computers. The manufacturers will tend to tell you they can locate those things anywhere and I wouldn't believe that. Normally if you have an environment that a control panel will stand up in so will a computer. If you have an environment where it's going to rust out in a year, so will a computer. Normally we like to see conditioned rooms for computer systems. And that can be like in Winnepeg we're doing a job, we're putting some of those little tin shacks you can buy and bringing in some outside air in a little air conditioner. But something to protect it from the environment. If you have to locate one on the floor near your process, there are cabinet air conditioners that you can install on those with filters but again they're not going to filter some of that corrosive air. And you put them out there, they're not going to last.

Lane Keffer: The corrosive air east our air conditioning units up. We have to replace them quite frequently.

Robert Skrentner: So it will eat your computer up too. You know you're just going to have to get used to it. You're going to have a lot of maintenance, I think, and a lot of replacement, or the best thing you could try if you've got instrument air around the plant you can try cabinet purging with the cabinet air conditioner and bring some good clean filtered air and purge your cabinets. Have those things put in a nema 12 or something and get a cabinet air conditioner and that will help with air purging.

Lane Keffer: We're roughly half a mile from the plant where the compressed air is.

Robert Skrentner: You have a problem. You know it's the corrosive atmoshphere in all plants that can raise havoc with any kind of equipment, be it analog or computers, and I don't think they're going to be reliable one way or another. You know those digital components are going to go on you.

Lane Keffer: In the effluent stream the pH drops below 4 sometimes and usually between 4.4 and 4.5

Robert Skrentner: You're not splashing your equipment with that, are you?

Lane Keffer: No, when it gets to the AB it is neutralized to about 7. It hangs around 7 most of the time.

Dave Keller: With our computer system, when we went to install it Foxboro said that you have to have an environment clean or a clean environment to put them in. So what we have done to keep any computers from failing on us due to corrosion is we put in big pace air makeup units and then running those in series with the air conditioning units. And it cleans it up very well and we haven't had any problems with it yet. But you have to have a dust-free environment for all your software and that's why we built the big computer control room and that also is run through the big air makeup units and air conditioning units in series.

Robert Davis: We had kind of a bad experience with the same thing because our original construction estimates did not include any of that, so we had to add in $55,000 of air conditioning/humidification which has never worked because our water's got about 750 parts hardness—can't get a humidifier to work under those conditions, and also putting a cage around the computer for static electricity and radio interference. Also putting a UPS system in it. So there's almost $75,000 or $80,000 of extra cost to the computer installation.

Dave Keller: There's one other point I'd like to make as far as computer installation and UPS—I would not purchase a computer without putting a UPS system in. If you crash, you're going to wipe out your program and it really creates a real havoc with your system.

Robert Skrentner: I have two comments related to two things that were said. One is on UPS. If you get distributive control systems and you're putting them out on the plant site UPS might get expensive and then I would go to the Farrow resident line conditioners—but something to condition the power. Just normal spikes in the plant will wipe those things out, too. So even if you're going to put one on a floor and air condition the cabinet or purge it or whatever, and if you don't on your main computer you should have a UPS and I'll agree 100% with that, but on the distributed ones you should also make sure your power is well conditioned 'cause that'd wipe them out. Also on environment, I was at the city of Milwaukee. They don't have computers, but they have multiplexers throughout the plant, and what they have are very large galleries and we were a little worried about putting some of the distributive controllers out there under an expansion and so we went out and inspected their multiplexers and they've been out there for about 5 years, I guess. They leave the doors open because they're nema 12 cabinets and if they close the doors they overheat, so the doors are all open and we inspected them and there's very little corrosion, at least in those galleries. Now they're up kind of on top but it's like next to the motor control centers, okay. The environment is good enough that in that particular application they don't apparently get that much in the way of corrosive gases. So I think there are areas of the plant you might be able to locate computer stuff and not get into too much trouble even if it's just at ambient temperature, but if you've got a lot of chlorine or pickle acid you can see what that does to plants, any of that kind of stuff get it as far away as possible and put it in a nice environment if you can. Otherwise I'll give you a couple years to run maybe and then you can start replacing things.

Dr. Robert M. Arthur: Any other questions on that topic—computers? I don't know if I should ask this or not, but since Smokey and I are the same generation, you didn't say too much about computers in your little talk there, Smokey, about how you utilize computers in a process control scheme. You talked more about the parallel flows in tanks and things like that. What's your impression of computer use?

Vernon Stack: I would say that I think computers are beginning to move towards the point where their overall utility is practical. I think a lot of the applications are either fairly sophisticated municipalities, I did say municipalities could, you know, large municipalities probably can do it pretty well. Smaller ones usually won't have the personnel to really make the thing tick. And then underlying all of this, which was mentioned in the talks, is the fact that if they're going to be practical the inputs, the sensors, must really bring in the information. I'm not still convinced that we have good sensors around to make a whole system work continually without having a whale of a lot of instrumentation experts or the kind of people who are around keeping the thing ticking. The sensors I think are coming, since your areas are coming in they're improving. But I'll be happier when there's more well designed and fully capable sensors to make the inputs for control purposes.

Dr. Robert M. Arthur: Ed, you haven't said a word. You let him do all the talking this afternoon. Any other questions?

Sigitas Viskantas: I've got a question on troubleshooting in control parameters if I can get back to that. I've got a problem in my final effluent in that I have a high filterable BOD of around 20 which normally you should have around 5 or so. But at the same time my volatile suspended solids is less than 10. I was wondering what parameters can I look at possibly to change to correct that problem?

Vernon Stack: Could you add to the picture, what's your influent BOD?

Sigitas Viskantas: About 250.

Vernon Stack: About 250. Is this industrial or domestic?

Sigitas Viskantas: No, this is domestic sewage. I don't have a respirometer either.

Vernon Stack: You need a respirometer first. But I think the things that you need to know, you have a high soluble BOD, you wish to get that down, so it seems that as far as the basic process design is concerned you don't have enough operating time, or your kinetic rate of operation, which the bugs are operating, is too slow and the effluent is getting out before it's treated to the degree that you want. so I say that's your problem. Exactly what's creating the problem? Do you have a slow kinetic rate because of some factor which is in the wastewater which is causing it, or has the process been designed without enough reaction time under the kinetics that you set up—under the conditions that you set up?

Sigitas Viskantas: Well, what should I look at? The respiration rate at different stages—primary and the activated sludge—and see what I can correlate, or to see if that is the problem of . . .? Prior to that the problem was reversed, where I had low dissolved BOD and my suspended solids was high. And that was due to backlog of solids that couldn't get out of the system. Now I'm in a different operation as far as land fillings, so there's no backup in the solids.

Vernon Stack: The kinetic rate that you're going to have in this system is going to be, well, let's say for example, let's suppose that you could decrease your F/M practically and you could increase biomass in the system, then you would be moving the kinetics towards providing a lower soluble BOD in the effluent. Is that possible in your system now, you said you had too many solids, you got rid of it? Did you get rid of too many, do you have too few solids in the system now? Too little biomass?

Sigitas Viskantas: I might have. That's where I think could be the problem.

Vernon Stack: Do you have limitations to being able to increase the biomass? Do you have secondary clarifying limitations on the return or return rate limitations? Any reasons why you can't bring up the level of biomass?

Sigitas Viskantas: Well, right now we're returning it at 60%, we can go as high as 100% returned.

Vernon Stack: Well, depending on what the initial settling velocities in your clarifier underflow rate are, going to 150 may not bring you back more sludge. You may just get thinner sludge. So that's one thing you have to understand is whether or not you have enough clarifier underflow area to get back the biomass that you need. It's a possible limitation in your design, but it sounds like you need to have a higher concentration of biomass for the inflow concentration so that your rate would go higher and your effluent soluble BOD would go down.

Sigitas Viskantas: Okay, thank you.

Robert Skrentner: I have a question on training, if I can switch subjects. Again, how many is most everybody here still from users or owners, operating agencies? I was curious. I think I talked to Walter at lunch and he said you estimate what—$80,000 per operator for training costs? Is that a real number? Anyway, I was looking at what I think the Commission in St. Paul estimated about $25,000 or $30,000 per operator in training costs, they thought that was reasonable, and I'd just like to get some thought on cost per operator in training. What's a number that should be budgeted for in that sort of thing?

Ed Fioroni: I'm glad you asked that question. I was going to make a comment by the way, Bob. Talking to some of you the comment that came out most consistent was you must have spent a fortune on what you did at Polysar. And that's what it looks like I think, but when we sat down and we looked at the $26 million project the total training package that you saw this morning in the presentation was approximately 1% of that. So that adds out to about $30,000–$35,000 per operatoer over a period of 5 years. So it's not that great if you consider the expenses that will derive from not training at all, and I think we had a real good example in Canada. I'm not going to name the company. They spent—in Montreal as a matter of fact—they spent about $28 million in designing a real good plant similar to ours, but did not take care in spending any money in training, and I think that their first 3 years they spend $3½ million in training and consulting to get the plant going. So that's why we put on good training program—but it's really not that much when you look at it for operators.

Dr. Robert M. Arthur: Does anybody know what happened in Detroit, because I know that they've had problems there with training. Does anybody know anything about that situation?

Pat Oates: I participated in many of the training programs at Detroit, and what it eventually developed into was free sleep time for the operators. Another question I wanted to direct to the gentlemen from Polysar, do you have any operator resistance in changing the picture of the operator from just somebody who's just out working with pipe wrenches, and you know big hammers and stuff, to getting them into the finer control process. With our operators they seem to want to stick to their narrow view of them being out in the plant and they don't really want to think.

Ed Fioroni: That's a good question because we did have the problem at the beginning but we were very fortunate to hire first of all three experienced operators from outside and then the other three operators—the Big O's we call them—the operators that made up the other six, we interviewed over 150 from within the plant. So we were very selective in choosing probably the three top operators in Polysar so that 6 unit really made up a very strong start and then it just led from one to the next and the other 6 assistant operators got to know what was going on. Of course we had a very open management in the biox which is a bit different than many other units. We had a very strong quality of work life which is operator-oriented and it attracted the other 6 and again we had almost 200 apply for that. So for startup purpose we went weeding out. Now, what will happen next you're probably right. We may have some problems in the next bunch of operators that come in, but hopefully these will have set the pace and then you know it will get out to everybody else that we are a bit different.

Pat Oates: But we're having trouble getting away from the "get the bigger hammer" idea. You know if it doesn't work, "get a bigger hammer" and the operators are more of the idea "tell me what to do and I'll go do it" but they don't want to make any of the decisions themselves.

Walter Frais: The one big advantage we had, we hired them and we scared the hell out of them before we hired them. In other words, if they didn't want to work we told them you're going to be miserable with us. So we got the people we wanted. And we'll do the same thing if people leave us. We're going to tell them if you want this job, you're going to be doing a lot of studying, you're going to go to school on your own time—we'll pay for it, and if they're not interested in that, they're not going to be working at our plant. It's as simple as that.

Sigitas Viskantas: I've got a question for Polysar. How many people do you have or do you have any that's full-time devoted to training and going and testing and monitoring as this training is being progressed?

Ed Fioroni: Basically, we don't have anybody full-time devoted to training. We have an individual, an operator, which we've given most of the responsibility to develop the laboratory training which we felt was a most important part, and the other part was developed by myself and Walter like the four-phase training program that you saw. All the laboratory work was done mainly by one operator and the rest was done by myself, Walter and the foreman in the last year.

Sigitas Viskantas: Yes, but to continue on, when you gave a person coming in you had all these different maths and somebody has to sit down and monitor his progress and or retests of some sort and seeing how he's going and that takes time and effort whether he's an operator, superintendent, or somebody's got to take time and do that. You don't have anybody, really. You've just got this regular staff and no additional person is what I'm asking.

Ed Fioroni: Yes, we don't have anybody at the moment. Once the program is developed I have it in my budget to get a trainer for the department.

Robert Davis: That's real nice luxury to be able to pick your own people. I know when I started with the Sanitary District and we were starting out the advanced treatment and computer facilities, and probably half a dozen people with an eighth grade education who'd done nothing but pump sludge for the last 20 years of their life. It's a lot more difficult to get them acclimated, especially when you're going to a computerized system. They do seem to work with it now. I mean they do interface with the day to day, they still kind of resist making process control changes and really want to leave that to the supervisors.

Dr. Robert M. Arthur: It's really strange that the plants I know all around the country and how the training and motivation changes from plant to plant. In the city of Milwaukee you have two different plants—Jones Island and the South Shore Plant—and you know what I'm going to talk about, Bob, cause I think you mentioned this. There's such a vast difference between those two plants because the South Shore Plant is motivated to experiment and to innovate and to use computers, but the Jones Island is not at all. Maybe it's because of the size of the plant and I can appreciate what you've been going through in Detroit, but the other situation I know of Green Bay, Wisconsin where they bring their people in and they rotate them. It's about a year or two training program and Terry Charles you went through that, right? And that's a good program and they—now he left, Green Bay, maybe he can tell me why—but they don't lose too many people, but the people they do lose go on to pretty good jobs because of the training they've had, and I've always felt that type of scheduled training and rotating them into the spirit that they're doing something. I'd like to hear some comments from you Terry because you went through that.

Terry Charles: That plant has been running for about 7 years now, 7 or 8 years. I started there in '76. It was a great training experience. I started on the bottom there and the training program started just as the Polysar people have indicated ahead of time with the skeleton crew and then beef them up as the process started up and came on-line. They rotate everybody through the system and I guess the encouraging is that the startup wasn't as smooth as it looks on paper. But the nice part about it was most of the troubleshooting was done by the hourly people out in the field. Once the vendors, once the acceptance tests were completed, then the problem started, and a year later there was some tremendous amount of teamwork and cooperation and downright plain thinking by the operators to troubleshoot some of these sophisticated pieces of equipment and a lot of the answers were very simple, yet the experts a year prior to that couldn't help us. So there they do try to promote individual thinking in each process and as you spend your 3 months, 6 months, depending on the level of operator you're at, you learn the process from the ground up and develop a feel for it and once you leave it you know you're coming back a year or so later so you try to retain that knowledge. They also keep a pretty good communication program amongst the people out in the field that keeps everybody informed on what's going on in all three sections so when they come back 6 months later, they don't have to start from scratch and retrain some of these people. So the in plant communication is very important. And I'm talking larger plants here where you have an operator team of say 15–20 people you must communicate from one end to the other and that doesn't mean hop on your golf cart and go say hi and have a cigarette with them. Something formal, some type of formal report must be passed around for each individual operator to see and read every week to keep them up to date on what's going on in each section.

Ed Fioroni: I just want to agree with what you just said. I think the key with us was to have, this is going back to Detroit there, was to have the operators onsite one year in advance because we had a lot of problems, you know it sounds rosy, but we had people walking out on courses when we sent them just for that course, they're not used to it. They weren't going to go out and do extra work. We had I think 5 or 6 walk out of one course because they weren't going to do it so we had to sit down and talk to them and make them understand that if we didn't have that one year, we probably wouldn't have been as successful as I think we will be. So it really matters that you get them in on the groundwork and develop with them. The pilot plant really helped us.

Dave Keller: I'm going to direct this to the Detroit people. I believe you people Detroit has Kompresses?

Pat Oates: They weren't online when I worked there. They were being installed.

Dave Keller: What I was going to ask was what your experience was with them, but if they weren't online, you can't tell me. I've got two of them, but my experience isn't all that great with them.

Tom Krueger: We've got the Andritz belt press and we've had a lot of good success with it and we're real pleased with its operation. Presently we're in an expansion program at our plant—construction is underway now and we speced Andritz again because we're upgrading, increasing the size of our dewatering operation, so we're looking at a larger belt press and we speced Andritz, Winklepress, the people who manufacture that, I don't remember the name, and Filtack which is—it's another type of belt press very similar to the Andritz. Filtack piloted our plant and that was one of the conditions at which we allowed that then to be speced. And it's very similar to the Andritz press and it's the one that got the low bid and so I don't know how that's going to operate but we're hoping it's going to be as good if not better than our Andritz.

APPENDIX

QUESTION AND ANSWER SESSIONS

CHAPTER 1

QUESTIONS AND ANSWERS

Owen Boe: Process control is trying to control effluent qualities. If—this may just show my weakness in statistics—but with that analysis—if you redefine process control as the objective of controlling sludge quality—filaments or SVI as an example—would those data then not show that if your strategy was changing your SRT (to change that sludge quality), that has no correlation to effluent quality or would they support that you are controlling more of a sludge quality type parameter and that would be because of a negative/positive interpretation to effluent control strategy?

Dr. Edward Schroeder: Ah, I may have to think for a while on that subject. No, I don't think so.

Owen Boe: I guess the reason for my question is that it seems like it's come up in some of my work that we've looked at and show that there is no correlation to SRT and effluent quality and that's a no control program and it seems for a person who's out there day to day operating a plant, he changes his SRT to meet SVI objectives, that support that he is having a control strategy there, and it is working and he's got good effluent quality as a result of his control strategy.

Dr. Edward Schroeder: Yes, in a sense that's right. Yes, in a sense that's right. I think that what I said was that within the range of values you use for SRT, for example, you don't have any effect on effluent quality. And if you want to talk about the affect of SRT on filamentous bulking for example—it's true, I can demonstrate the thing but I can also demonstrate the complete type of thing. If I have a very long SRT, I can produce bulking; if I have a very short SRT, I can also produce bulking, but it would be a different type of bulking. You do have some direct control with SRT over bulking and I don't think it shows up. I think we're talking about apples and oranges.

Walter Schuk: I think you should go right up the coast a little ways and take a look at Hillsborough, Oregon when you said that you can't find any place where they changed the operation of the plant to control the plant. That's one example that I know of. It's also written up well enough that it's done with a programmable pocket calculator. They simply take their measures when they do change plug flow, multiple-stage, step aeration and contact stabilization. And the other thing I'd just like to add a very brief comment sometime to you today on instead of taking 9000 numbers that are produced by daily grab samples from 83 plants, how about doing some of this with automation and see what's going on on a day-to-day basis. It might change your statistics a little.

Dr. Edward Schroeder: Second question first. Quite possibly, these were not grab samples but a composite. But on the first question, no I didn't say that plants don't do this. In fact, taking the major share of plants around my area of Davis where they operate in a flexible mode of operation and that they operate usually 3 possible modes—conventional plug flow, nominal plug flow, step aeration and contact stabilization. What I said I think that's true and I don't think it works. It's not satisfactory.

CHAPTER 2

QUESTIONS AND ANSWERS

Walter Schuk: Bob, you covered a lot of points there, but there's one thing I think you left out and that's time in the aerator. And that is controllable by recycle. Got any thoughts on that?

Dr. Robert M. Arthur: Yes. Very good, Walt. What do you do if you try to increase the microorganism population—I hope this is what you're talking about—by increasing the rate of return sludge? You hope that you're getting more viable microorganisms in there, but you're doing something far more serious and that is you are reducing the detention time in the aeration tank—hydraulically. Now, if your treatment time in the aeration tank is not equal to or less than your detention time, you're in trouble because unmetabolized food goes out of that aeration tank, so that you could be doing the worst thing possible by increasing the rate of return sludge because you decrease the detention time in the system. So that has very serious consequenses. Is that what you meant?

Dr. Edward Schroeder: Not to complain, Bob, but since my name was brought up I thought I'd just make a point. The definitions of F/M and SRT and so on that I use are based upon standard definitions. You've modified those definitions a little bit so it's a little bit unfair to say you can't derive those two separately. And in fact, they don't depend upon BOD_5 or MLVSS or anything—the relationships only depend on mass balances that you have to accept as being commonly defined. So if you change the definition, then you get a little different relationship, that's all.

Dr. Robert M. Arthur: But the things that are used do not have biological significance because SRT is a physical measurment, right?

Dr. Edward Schroeder: Well, it comes right out of a mass balance in the kinetic relationship.

Dr. Robert M. Arthur: Of Solids?

Dr. Edward Schroeder: No, in fact . .

Dr. Robert M. Arthur: Of what?

Dr. Edward Schroeder: You get the relationship you're interested in by making a mass balance on the solids around a reactor, but what you end up is eliminating the solids from the system and you end up with just a relationship between SRT and effluent organic concentration, and it just falls out. You can't avoid it—in fact, there has to be a single valued relationship between the two. You get the relationship between that and F/M.

Dr. Robert M. Arthur: But you're trying to show that what you claim is a biological measurement of BOD_5 . .

Dr. Edward Schroeder: Well, I don't use BOD_5 ever. I always use ultimate BOD or TOC, it doesn't make a difference.

Dr. Robert M. Arthur: Well it's still not a measure of the exact food that's in a system.

Dr. Edward Schroeder: Sure it is. I would just really strongly disagree with you.

Dr. Robert M. Arthur: That's good. We have some disagreement.

CHAPTER 5

QUESTIONS AND ANSWERS

Walter Schuk: Went through this with one of your people, try a new guy and see if he gives me a different answer. Where do you put dollars in your report structure? It's one of the most useful tools the manager has in prying money out of his management to get your process to work.

Owen Boe: You're absolutely right, especially in a private organization, that our managers are graded considering dollars control. We don't yet— I think it has to be a building structure first of all and the next level that we're working on is to zero in to more on what—these were process control parameters—is a process in control or out of control? To balance that you need to consider the consumption criteria. What is that decision or what decisions made in that process impact in terms of consumables? So we just have finished a prototype of a program that then complements this and zeros in onto the consumables. And by the way, you can report any parameters on the system any way you want, but in terms I think of where your question was heading you need to balance a pure process decision with a consumable type comment and I would just offer though that that's not necessarily a very straightforward solution. Some things are very straightforward about it. Pounds of chlorine per million as an example. Obviously when that number goes up your dollars go up. When you make a process decision as to change that target or change something about that you just have to be conscious that you may be taking the balloon effect and squeezing the balloon and having it pop out somewhere else. So that consideration has to be factored in there somewhere.

Pat Quinn: How many parameters do you recommend for a plant that you can keep track of?

Owen Boe: For a full data base?

Pat Quinn: Yes, for an activated sludge type plant.

Owen Boe: I'll skirt that briefly, because I'll say in terms of the top-down approach I think two or three is from the plant manager's standpoint plenty adequate. If I was process control director I may end up with on a routine basis, and when I say routine, I mean when I want it on a flash report and I want to put that much emphasis on it I want it to be a very regularly monitored parameter. From a process control report it may be a dozen, maybe 15, depending on the size of the plant, the complexity of the system, and after that there's a lot of background information you can add to it that's nice to know and depends again on a lot of circumstances. I try to avoid especially when working with our plants saying you *will* control this with an F/M strategy or respiration rate strategy. There's a lot of different strategies so we go through and at that level decide what is it that you want to control this process now, uniprocess, and what data do you need for that? But generally a dozen parameters pretty well sums up an overview, but keep in mind that that's backed up with probably a lot of other data too.

Dr. Robert M. Arthur: Can you give me a brief answer on what you're talking about in terms of dollars for a system including software, and is Envirotech marketing this system?

Owen Boe: That's a leading question if I ever heard one. Yes, we're marketing this system now. In a sense we're not in the computer vending business so we don't sell our software per se; we don't sell computers per se. We have a concept, we have a package, we have a program. So we deal with clients on a case-by-case basis of just providing a package system to support a structure that they may or may not want for their plant, and just approach it on that basis. I avoided the question. I can't tell you that. We don't sell the software, no. But we can put the complete system in a plant, and so it's there so essentially on a lease type basis or setup basis so it's there for you to use as long as you want it. You know, there it is. Kind of like the video game, Xerox machine type things and then when the upgrades come along like the energy consumable concept comes along that becomes an upgrade that is fixed in the package.

CHAPTER 8

QUESTIONS AND ANSWERS

Dave Breuer: I work in an industrial situation where periodically there are wastes present in the waste stream which take quite a while for the respirometer to pick up—there's a long time lag, some acclimation period. It can be as long as 30 to 60 minutes before the biology in the respirometer really starts to assimilate that waste. Sometimes these things are present and sometimes they're not. If I were to try to correlate my waste stream strength with respirometry over a short period of time, I would think that I might miss those wastes, they would never start to show up in the test. Could you comment on that and perhaps recomment a way around it?

Dr. Normand Therien: When you're doing that measurement, you're probably using already acclimated sludge coming from your process.

Dave Breuer: Yes, that's right.

Dr. Normand Therien: Okay. These sludge's are coming from probably your sludge recirculation stream.

Dave Breur: We've generally been using mixed liquor suspended solids.

Dr. Normand Therien: That may be . . . an interesting experiment to do here would be to use your sludge coming from your sludge recirculation stream. Make sure that they are in endogenous phase. All the experiments we've done here, surprisingly enough, were taking the sludge from, as I said the sludge recirculating stream, and we had to wait some time between 8 and 15 hours before we reached the endogenous stage, and when I was talking about a few months getting accustomed to the thing is simply because we assume that they were in the endogenous phase and then if you do that what you find out is that sometimes sludge coming from your aeration basin already has enough food in it—they store such a quantity—that before they get rid of it and start working on your stuff, it takes time. So it's not really acclimation time, it's really time to get rid of what they got some time already inside and we were able to observe that using very biodegradable substrate. You use beef extract and you see it disappear from the solution. They are just stored in the cell, but if you use these cells to then do another test, you find out that they don't seem to respond. They're just digesting. So what I would suggest is really do a test using some that are really in a starving mode and see how they compare with the type of result. Maybe this will disappear.

Dave Breuer: Okay, thank you.

Vernon Stack: More in the form of a comment on the industrial problem. Industrial waste can be so complex that in some situations when I have seen more than one where the very complex materials may be stored in the biomass, but for the practical sense of getting that to endogenous, nothing happens to that material until you put in some available energy. The moment you put in some available energy, for example you may add a small amount of available energy and then get a very large uptake for a period of time which represents that energy being available to handle some of those materials that are not handleable because they don't have the energy or there is some blockage that the energy breaks up. So we talk about domestic sewage that's one thing, but you talk about simple beef extract or other things, that's something else, but industrial can have quite a lot of complexity to it. You may find that the only way to get a meaningful value is to take the total area under the respiration curve which is energy oxygen, and this storage I've just talked about can foul that up so that it can get to be complex enough so that you don't—you can't—get the kind of handles which were developed. These are very convenient handles. They are taking a piece of the total the same way BOD_5 is taking a piece of the total and correlating and if it's too complex you may have to look at the total. It's like where you would have to run a total ultimate BOD to get information, you may essentially have to do the same thing with the industrial wastes.

Dr. Normand Therien: Thank you for your comment. One comment I have to do here, and I have to excuse myself. Slides I've shown here, and some of the slides may have amused you when looking at the ordinate because sometimes you had respiration rate there. They were oxygen demand, they were integrated values and maybe have amused some of you in error when you look at the slide—they're all correct in the paper. All of the ordinates were really integrated values over 30 minutes. They were not respiration rate. One thing also, I don't want to leave the impression that we solved the overall problems when I said here's an example with industrial wastewater. We're working with some of the type of industrial waste which shows toxic behavior and probably next year I'm going to talk about that. We're finding that the respirometer will also give you information on how to treat wastes also.

CHAPTER 10

QUESTIONS AND ANSWERS

Dave Keller: I'm wondering about your operators. When you put your operators on, you say about a year ahead of time, were these operators in the wastewater field at all prior to this time?

Ed Fioroni: Of the twelve operators that we have, well we have six operators and six operator assistants, of the six operators, our key operators, three had anywhere between 6 and 20 years wastewater experience and were all university graduates, the other three, no. They were very experienced operators within the plant but had no wastewater experience. And the other six had none.

Dick Meagher: It seems you put quite a dollar investment into this training. I didn't quite understand, how did you ensure that this operator would stay at an operator's level and not move up to the plant?

Ed Fioroni: How do we assure that the operator remains at the operator's level?

Dick Meagher: And not move on through the organization.

Ed Fioroni: Oh, okay. That's a good question. We couldn't ensure that. We're assuming that they will find this job because we do put lot of time and effort into it and we are ensuring a job permanancy for them that they will consider this job as a permanent one for them, but they can post out any time if they wanted to. But hopefully we'll have a solid training program established so that anybody that comes can be quickly put into the spot that he was in. But we can't keep them forever. No, if they want to move on, they can.

Bob Davis: My question is pretty much the same in that I would be afraid that after a 5-year training course to that extent that the people would be leaving to manage a plant like MSC Chicago or something. You're giving them an equivalent of a masters degree almost and I would think that the cost of training these people would be prohibitive and you probably end up losing 80% of them.

Ed Fioroni: Okay, the cost of the training program came out to about 1% of the total cost of the plant. And that was my main point with my management that I felt that it was necessary. Yes you're right, we are giving them an extensive training, but I feel strongly that without it in a plant like Polysar, which is very strongly union oriented, we had to go over and above what most plants did. And if we hadn't done that, we would be in serious problems.

Bob Skrentner: I really don't have a question, but perhaps a response to the preceding question. At the Metropolitan plant in St. Paul they spent an awful lot of money training operators also, and they have computer simulators and they had them all running around the plant and spent about 2 or 3 years with their control room operators. And they've been running for about 2 years now and I think they've had one operator request a transfer back to the field and that's the only one who quit. They're quite knowledgeable; I think it's because of the job satisfaction. They don't feel lost and panic and so far they're doing real well with them.

Dr. Robert M. Arthur: Is that a municipal?

Robert Skrentner: Yes, that's a municipal plant.

Ed Fioroni: This is industrial and I'd just like to make a point that in an industrial plant like ours these operators are paid at the same level as anybody else. They're not paid more nor less. So we have to make our job description a satisfactory one for them to stay with us.

Lane Keffer: I'd like to address the design of the plant. You said that the On-Line Respirometer would tell you when you had a toxic material coming down the pike. How do you plan to get rid of that toxic material out of that stream, or how do you plan to deal with it?

Ed Fioroni: We expected that question. In Ontario right now we have a deal with the Minister to the Environment that they opted out to let us bypass if there's any problem. So what I'm saying is that if we had a choice to destroy our bug population and bypass to the river, they would accept that for the next 2 years. So we can bypass when we have a problem. Which is nice to have.

Sigitas Viskantas: I realize this is an activated sludge conference, but part of our problem is what do you do with the sludge that you do generate?

Ed Fioroni: The other question we expected. At this point in time we don't have any facility in design what we're going to do with our sludge. Again, we're just going to put it to the river. And it's a matter of seeing what the Ministry will come up in the next couple of years, and I'm sure that if we have enough sludge, we will have to devote our attention to it and solve the problem. But right at this moment we don't have anything to do with sludge.

INDEX

activated sludge
  process variability 54
  purpose 32
aeration, control of 32
anticipatory control of aeration 39, 36,104

biological phenomena 26
BOD—relation to respiration 113, 123,181

cellular
  control 28
  metabolism 28
chemical addition 7
coefficient of reliability (COR) 18
computer systems
  specifications and installation 78
  maintenance 83,198
control
  of aeration 32
  anticipatory 39,63,104
  of dissolved oxygen 6,104
  of return sludge 33,40,101
  of solids distribution 95
  of wasting 35,103
controllable variables 32
control strategies 37
  application 38
  definition 37
  limitations 43
control tests 48

data management 66
  bottom up 66
  top down 67
  microcomputers in 72,74

data selection 55
data utilization 57
design, facility 135
dissolved oxygen 6,104

effluent quality, variation 16

facility design 135
flow equalization 4
food (biological) to microorganisms (viable) ratio F_b/M_v 40,62
food/microorganism (F/M) ratio 5,35
food monitoring 29

instrumentation 148
  on-line and laboratory 158

leading process indicators 69
loading rate 5

management concepts 65
mathematical models 17
mean cell residence time 2 (MCRT) 49
microcomputers
  in data management 72,74
microorganism storage 36,192
microscope exam 38,49
monod equation 3

nitrification and respiration rate 191

operator training 163,203
  start up 166
  pre start up 163
  post start up 167

oxidation reduction potential 42
    (ORP)

physical phenomena 29
process configuration 6
process control 179
  by computer 80,81
  definition 2
  F/M 8
  job description 72
  objectives 2
  organization 71
  parameters 57
  plan 69
  practice of 89,195
  SRT 8
  in start up 192
  support systems 47
  techniques 78
process control strategy definition 26
process controlability 154
process flexibility 135
process operation 154
process performance variation 1

reactive control 63

recycle ratio 6
respiration 50
respirometers
  to determine BOD 113,123,181
  implication for control 124
  biomass related to MLVSS 114,118
respirometry 40
  on-line 37

sedimentation 29
sensors
  in process control 84
settleometer 49
settling velocity 30
sludge age 2
sludge quality in control 48
solids inventory 50
solids retention time (SRT) 2
statistical tools 18
storage of
  raw wastewater 36
  microorganisms 36

time scale response 17
toxic loads determination 160
treatment process instability 16